A Victorian World
of Science

A Desire of amufement, and relaxation is natural to man. The mind is foon fatigued with contemplating the moft fublime truths, or the moft refined fpeculations, while thefe are addreffed only to the underftanding. In philofophy, as in polite literature, we muft, to pleafe and fecure attention, fometimes addrefs ourfelves to the imagination or to the paffions, and thus combine the *agreeable* with the *ufeful*.

For want of this combination, we find that pure mathematics (Comprehending *arithmetic, geometry, algebra, fluxions, &c*) notwithftanding their great and acknowledged utility, are ftudied by but few; while the more attractive fciences of experimental philofophy and chemiftry, are almoft univerfally admired, and feldom fail to draw crowds of hearers or fpectators to the lectures of their profeffors . . .

Encyclopaedia Britannica, XVIII
Fourth edition, 1810

A Victorian World of Science

*a collection of unusual items and anecdotes
connected with ideas about science and its applications
in Victorian times*

Alan Sutton

*Department of Education,
University College, Cardiff*

Adam Hilger Ltd, Bristol and Boston

This selection and arrangement © Adam Hilger Ltd 1986

British Library Cataloguing in Publication Data

Sutton, Alan
   A Victorian world of science: a collection of unusual items and anecdotes connected with ideas about science and its applications in Victorian times.
   1. Technology—History—19th century—Sources
   I. Title
   609'.034   T19

ISBN 0-85274-559-1

Published by Adam Hilger Ltd
Techno House, Redcliffe Way, Bristol BS1 6NX, England
PO Box 230, Accord, MA 02018, USA

Typeset and printed in Great Britain at The Bath Press, Avon

Contents

PREFACE   *pages ix–xi*

INTRODUCTION   *pages 1–8*

CHAPTER 1
PRACTICAL SCIENCE and the USEFUL ARTS

Education and industry; Working men in want of a living; Deception in the weaving trade; Technology and unemployment; Vivonnaise water wheel; Steam battery; Hydroelectricity; Water motor; Liquid air as a power source; Refining crude oil; Alcohol from acetylene; Medicine for cows; Wind motors; Heating for factories; Safety of flywheels; Boiler explosions; Hints to mechanics; Removing rust; Burnt steel; Roquefort cheese; Preservation of meat; Expanding cylinder engine; Bourdon engine; Wagenknecht rotary engine; Melhuish hot-air engine; Transmission gearing for tramcars; Concrete cottages; Bridge wrecking; Alternative geomagnetic field; Power to move the earth; Tesla's torpedo; Wireless telegraphy; Head-rest for field glasses; Electric pen; Congestion on stairways; Hints for holiday makers; Holiday craze; Lorgnette humaine; Dangerous advertisements; Paper cannon; Engineer's epitaph.

pages 9–56

CHAPTER 2
DOMESTIC SCIENCE and PRACTICAL HOUSEHOLD ECONOMIES

Washing machines; Cheap servants; Eliminating vermin; Blackbeetles; Burglar alarms; Telephone; Electric candle;

Incandescent mantle; Illuminating power of gas; Air pollution by gas heating; Gas-powered alarm; Noisy neighbours; Electric cradle; Heat motor; Solar energy; French bath heater; Universal Teapot; Water closet; Turnip bread; Adulteration of foodstuffs; Wholemeal bread; Salt in the household; Cleaning skulls; Potato soup; How to cook salmon; Roast lamb; Transporting butter; Waterproof garments; Wool versus linen; Varnishing chairs; Care of shoes; Furniture polish; Removing lead from carpet; Cleaning goatskin rugs; Cosmetics; Worming the cat; Tea and coffee making; Home-made wine; Teleanimatograph; Cleaning rubber shoes; Preserving green peas.

pages 57–95

CHAPTER 3
PHILOSOPHICAL AMUSEMENTS, PASTIMES and HOBBIES

Experiments for entertainment; Electromagnetic induction; Telescope; Microscope; Camera; Exposure meter for pin-hole camera; Viviscope; Praxinoscope; Optical theatre; Magneto-telephonograph; Phonoscope; Archimedean screw; Aerodynamic top; Rod and ring experiment; Boxing kangaroo; Electric motor; Thermo-electric motor; Self-acting fountain; Universal sundial; Automatic parrot teacher; Puzzles and games; Three-handed draughts; Balloon bicycle; Sailing on skates; A lecturer's experience; Answers to puzzles.

pages 97–134

CHAPTER 4
MEDICAL SCIENCE

Life and death; Care of teeth; Fillings for teeth; Constipation; Electric bath; Electric douche; Electric shock; Therapeutic value of creosote and anthracite; Explosives we swallow; Headache; Refrigeration as a cure; Castor oil for congestion of the brain; Loss of hair; Sources of infection; Malaria; Therapeutic value of venomous secretions; Earthworms as a cause of disease; Vaccination; Causes of disease; Tetanus cured by tobacco; Smoking and heart disease; Prevention of aging; Omnibuses as a source of illness; Varicose veins; Corns; Ingrowing toenail; Shape of feet; Artificial foot; Artificial arm and hand; Amputation by electric saw; Artificial eye; Bifocal lenses;

Seeing competition; Reducing unnecessary fat; Photographing inside the stomach; X-rays; Stooping shoulders; Treatment of wounds; Doctor starves; Idiot genius; Invalid's piano; Hysterical blindness.

pages 135–179

CHAPTER 5
VELOCIPEDES and FLYING MACHINES

Newton's steam carriage; Locomotives on Highways Act; Michaux boneshaker; Phantom bicycle; Double-driver boneshaker; Power multiplier; Eureka driving wheel; Windmill velocipede attachment; Thompson's propelling mechanism; Eucycledian bicycle accelerator; Unicycle; Double-driver bicycle; Orthobaton bicycle; Cyclists' backrest; Clockwork tricycle; Joule's tricycle; Ratchet drive; Safety saddle-brake; Excelsior velocipede; High Peak velocipede; Ball-bearings; Rustless bicycle; Ayrton and Perry's electric velocipede; Gas-fired steam tricycle; Stanley steam carriage; Grindstone-ignition engine; Petrol-assisted bicycle; Reid's motorcar; To Coventry on a motor-car; Speed-measuring camera; Aerial Transit Company; Artingstall's flying machine; Kaufmann's aerial locomotive; Hall's helicopter; Hite's airship; Sky cycle; Langley's aerodrome; The state of aeronautics; Andrew Scott's prediction.

pages 181–227

To Dorothy and Clare

Preface

This book is the product of a random collection of items culled from the pages of a popular nineteenth-century periodical called the *English Mechanic and World of Science*. That magazine began publication simply as the *English Mechanic* in March 1865 and continued in weekly issues in one form or another until well into the present century before it was absorbed by other publications and lost its identity. During its early lifetime it too incorporated other magazines, including for example the *Mirror of Science and Art*, and the title was expanded to reflect the wider range of contents. The price also increased within a short space of time from one penny to two pence per weekly issue, at which level it remained at least until the First World War.

The early amalgamations vastly enriched the magazine and widened the readership. It was read not only in Britain but also in the then colonies and other countries throughout the world. In common with many nineteenth-century journals it extracted material from a whole variety of sources such as *Scientific American*, *Nature*, *The Lancet* and several other American and European publications. Consequently it became a convenient compendium of all kinds of information—reports of the activities of scientific societies such as the British Association for the Advancement of Science, news of scientific advances, articles on astronomy, chemistry, medicine, engineering and industry, electricity, telecommunications, transport, food and diet, scientific pastimes and puzzles, educational developments and much more. Indeed many correspondents to the *English Mechanic* claimed that they owed their education almost entirely to the articles it contained.

According to an editorial in the first issue, the magazine was aimed at '... the brain and pocket of the "bone and sinew" of the land, our great workers'. It must be said, however, that the majority of contributions from the readership to the magazine did not come from that quarter. Nevertheless, it was insisted—not without evidence—that '... there are thousands of working men who would dearly like to have under their hand at the fireside an epitome of the progress in the Arts and Sciences from week to week'.

Support came from letters to the Editor such as the following sent in by a reader from Batley in Yorkshire:

A WORKING MAN'S OPINION OF THE ENGLISH MECHANIC

Sir,—I have for some time been a subscriber to the ENGLISH MECHANIC, and am truly grateful to you for the publication of such a paper; I think that it is a proper publication for any one in my situation (I am an engine man) to spend a pleasant evening over after a hard day's work; the subjects are so varied, and so well handled by the various contributors, that I think, as a whole it deserves the highest praise the working mechanic or engineer can give it. In addition to its columns being open for correspondence by members of the various branches of industry, there is also the benefit of the 'Exchange' column for subscribers; and this offers peculiar advantages to the working classes, not to be found in any paper besides the ENGLISH MECHANIC. Hoping that your spirited journal may long continue to flourish, is the earnest wish of your obedient and humble servant—
RETRIEVER
Oct. 18. 1867

At this distance in time there is no knowing whether that was an editorial 'plant' but it does sum up the general atmosphere of the magazine. Although as the years went by there were evolutionary changes of style and presentation, the content and balance of articles stayed remarkably close to the original aim.

For me it is the contributions from ordinary people that make the *English Mechanic* especially interesting—whether in the form of short articles, or in answer to queries sent in by others. A reader, for example, might send in a graphic description of abdominal pains that he was experiencing (most correspondents were men). A week or two later he would be informed that he was probably suffering from appendicitis, and what he should do about it. Such exchanges are the foundation of this book. They give a colourful insight into the levels of understanding of scientific matters of ordinary men and women, and of the way that advances in science and technology affected their way of life.

One of the problems in putting the collection together was to decide what to leave out. In the end, the only practicable criterion for inclusion or exclusion was whether any individual item had any appeal for me. I make no apology for that, although I do hope that somewhere there are others who share my tastes. The vast majority of the items are quoted directly from the *English Mechanic* but in a few instances some editing has been done to make the items less tedious to read. The original sources have been cited where they are known.

Another problem was whether to translate 'old fashioned' units of measurement, where they occurred, into modern metric equivalents. To have changed the units would have destroyed much of the charm of the original items and to have added metric conversions in parenthesis would have been unnecessarily cumbersome. The problem was finally solved by ignoring it. Only in a very few instances are modern equivalents given.

It would not have been possible for me to put together this anthology without help. I am deeply grateful to Mr T Dawkes, formerly the librarian in the Department of Education at University College, Cardiff, to Mr G A C Dart, the former County Librarian for South Glamorgan and to his successor in that post, R Ieuan Edwards, for giving me access to rare copies of the *English Mechanic*. My thanks must also go to Mr Neville Hankins who encouraged me to put my rather haphazard collection of material into a form suitable for publication, and to all those who have helped me in so many ways to see it through to the end.

R A Sutton

Introduction

THE nineteenth century is a fascinating period. It saw the rise and gradual decline of Britain as 'the workshop of the World' and the beginnings of just about everything we take for granted today—even television (the 'Teleview' was a form of telegraph by which pictures were transmitted by electricity), aeroplanes and motor-cars. It is truly amazing to realise how many major achievements there were—railways, powered ships, the telegraph and telephone, electric power and lighting, photography, sound recording, x-rays, spectrum analysis, anaesthetics and antiseptic surgery to name only some of the most outstanding.

This book, however, is not concerned with the achievements of the 'big names' in science. They are well catalogued and discussed elsewhere. It is entirely an anthology of applications of science as seen by ordinary people—educated and uneducated alike. Some of the items are of an educational kind, others are homely, pragmatic applications of scientific principles, still others—which are especially appealing to this writer—are no more than speculations on how the laws of science might be bent to the service of mankind, whether propelling bicycles without effort or controlling the earth's magnetic field.

Some of the schemes proposed for achieving this or that end may seem ludicrous to us looking back from our present standpoint. We should not dismiss them out of hand but should try to view them in the context of the times and to some extent read between the lines to discover why people thought the way they did. Electricity, for example, was an endless source of wonder and inspiration. If anything could be done then electricity could do it—whether destroying bridges or curing toothache—if only the right way could be found.

For this writer, two things emerged from collecting and assembling the various items. One is a powerful impression of the cyclic nature of history. There is truly nothing new under the sun. The problems we face today such as unemployment, rapid advances in technology, economic recession, upheavals in education, pollution and the energy crisis, were all there a hundred or more years ago. The second is a vastly increased respect for the men and women who lived through those hard times and who showed tremendous resilience, enthusiasm and pride in being self-reliant—even to the extent of making their own artificial arms and legs to replace limbs lost in tragic accidents. It is surprising to find how many women were the victims of such misfortune and how accidents and injury seemed to have been accepted as inevitable. Where on earth did they find reserves of energy to walk miles to and from work, spend twelve hours or more six days a week if not seven on laborious tasks, and still find time for recreational pursuits? Their thirst for learning and self-improvement was phenomenal.

Towards the end of the century, voices were being raised in support of better and more meaningful education for the less well off, and at the same time criticising the consequences of the rigid examination system in much the same tones as we hear today. An editorial in the *English Mechanic* in March 1895 had this to say:

This has been called the age of examinations. And the coming part of the year is the time at which recur the annual exacerbations of this particular disease—that is as far as it affects the vast majority of the men and women for whom the ENGLISH MECHANIC especially works. At Oxford and Cambridge—but we emphatically put on one side Oxford and Cambridge. It is true that all the vast emoluments of those two universities have been produced by the working people of this country—the English mechanics. Every entrance scholarship, exhibition, and sizarship; every scholarship given as a result of the May examinations, or of special examinations, to 'members of the university', every fat fellowship, all the salaries of deans, 'heads' and chancellors, vice or otherwise, have been ultimately derived from the unpaid labour of the working classes. On the other hand, with the rarest possible exceptions, there is no chance of any member of that class enjoying the advantages of an Oxford or Cambridge education.

But, after all, under the present unjust conditions of education, the university of the working class is the Science and Art Department, South Kensington. By the 1st of April—i.e. next Monday, must be sent in to South Kensington all the drawings, paintings, modellings, designs from the art schools connected with the Peoples' University. And early in the month of May begin the long series of examinations in the 25 science subjects, ranging from mathematics to hygiene. To those examinations, it should be remembered by the private student, any one of either sex or any age can go . . . . If, however, any young man or woman reads this article who has been studying alone, and who desires to test by way of examination the result of work . . . quite free of any cost he will be able to submit himself for examination in such subject or subjects as he may desire. For the purpose of this last sentence, be it understood that wherever the word 'he' occurs, it shall be taken to mean 'he' or 'she'.

The article goes on to discuss some of the examples of examination questions and the effects of the examinations on syllabuses. One example cited was

Give an account of the researches of Lawes, Gilbert and Pugh, on the absorption of nitrogen from the air by vegetables, as published in the *Philosophical Transactions*, 1851.

To which one candidate is alleged to have replied

I was born late in November, 1851, and did not take in the *Philosophical Transactions* for that year. But I shall be glad to tell you what *I* know about the absorption of Nitrogen &c.

It was recorded that he passed!

The most remarkable thing about that article is the date. It could so easily have been written today rather than 90 years ago. Some things have changed but we are still arguing about the availability of university places, unfairnesses in the examination system, the qualities of examination questions and so on. Another point worth commenting on is the attention given to equal opportunities for women—again so topical today. One thing which, however, seems to have been neglected is the opportunity for *anyone* to turn up at a specified place and present himself or herself for examination in the expectation of at least gaining a certificate indicating meritorious achievement. Possibly the closest to that ideal in Britain today is the Open University.

Of course opening up examinations in such an unrestricted way could cause problems, as the following example shows. The question was on a London University matriculation paper in English Literature 'Write out the Lord's Prayer, and underline the words in it of classical origins'. The article records that the examination 'was known would be attended by Hindoos, Jews, Mohammedans, barbarians, Scythians, bond and free ....' A Hindu candidate was reported to have responded 'Not being a Christian I do not know the Lord's Prayer'. It is not recorded if he passed.

Another argument which raged at the time (1895) concerned the British Imperial System of weights and measures. Until I came across the following article I had not realised the intensity of the pressure to change over to the more rational metric system, a change even now not fully completed. The article has been abridged but, despite its length, it makes interesting reading.

The child's agony of learning tables is one of the most lasting memories of our early years. And as we grow up we see more and more how almost entirely unnecessary are the trouble and pain to which children are put by our barbaric and brutal system of weights and measures. If the boys and girls of this benighted country could only know what is going on at the present moment, they would send up from every board school, from every preparatory school for young gentlemen and seminary for young ladies, from every nursery ... an appealing and appalling wail to the British House of Commons; or, at all events to the Select Committee that has just been appointed to 'inquire if any, and what, changes in the present system of weights and measures should be adopted'. Sir Henry Roscoe, who, as a scientific man is *ex officio* against the English system and in favour of that of the rest of the civilised world, is chairman of the Committee.

At its first sitting evidence was given by Mr H J Chaney, superintendent of the Standards Department of the Board of Trade. This gentleman described the system under which the verification of legal standards of length and weight, or better, mass, is carried on by experts of the Board of Trade, and also described the want of system which obtains in English weights and measures. Only a few days before, Mr Chaney, together with Sir Courtenay Boyle, the Secretary of the Board of Trade, had received a visit from the President of the Royal Society and some five other representatives of that learned body. These scientific gentlemen had gone to the Standards Office in Old Palace-yard, Westminster, in order to see the new standards of length and mass recently placed there. Would it be possible in any other country than England for these standards to be those of the very system of weights and measures persistently ignored and rejected by the English people? And yet this is the case. The standards are metric ones.

For more than twenty years an international committee dealing with this international question has been at work. As far back as 1875, twenty nationalities promoted and carried out a metric convention. It is hardly necessary to say Britain was not one of them. She, after her historical wont, held sulkily aloof. And, again after the wont of history, she had nine years afterwards, in 1884, to tardily come in with the others.

And now let us think, although it will hardly bear thinking, upon the hours and days and weeks and months spent in trying to learn the abominable English system. It is no exaggeration to say that at least half, and probably more nearly nine tenths of the time spent by children over arithmetic could be saved by the adoption of the sensible metric system. Think of our 3 barleycorns (to be collected from the middle of the ear and carefully dried), 1 inch; 12 inches, 1 foot; 3 feet, one yard; $5\frac{1}{2}$ yards, 1 rod, pole or perch; 40 poles, 1 rood or furlong (hence the old riddle about the politeness of the Poles, of whom it takes 40 to make one rude); 8 furlongs, 1 mile. Note the bewildering, the agonising shifting of the numbers of the units—3, 12, 3, $5\frac{1}{2}$, 40, 8. And think how this confusion becomes worse confounded in the square measures of surface. Imagine the mental obfuscation of a child who has to learn by heart that $30\frac{1}{4}$ square rods make 1 square pole before he or she has even got so far in arithmetic as to understand that the square of $5\frac{1}{2}$ is $30\frac{1}{4}$. Then, our $2\frac{1}{4}$in., 1 nail; 4 nails, 1 quarter; 4 quarters, 1 yard; 5 quarters, 1 ell—a measurement by the way, connected with the word elbow, for, was not the ell the length of the King's arm? Our pints and quarts, and gallons and pecks, and bushels; our Troy weight for gold and silver; our Apothecaries' for drugs; our Avoirdupois for things generally. To say nothing of the different dimensions denoted by the same word. Or, to say nothing again of such odd weights as the carat, which in connection with gold means a ratio (e.g. the 22 carat gold of a sovereign means that 22 out of 24 parts are pure gold) and in connection with diamonds means a weight, $3\frac{1}{6}$ grains; or the butcher's stone of 8 pounds instead of the normal 14. Until quite recently a ton of stone was quite a different matter from a ton of everything else, so that the poor bewildered child might think there was something after all in the catch-question: 'Which is the heavier, a pound of feathers or a pound of lead?'.

The fact is that all this unnecessary time and trouble, both in working our own wicked system and in having to translate between it and the metric, are due to the intense obstinacy and pig-headedness which make us cleave fatuously to the Fahrenheit scale of temperature based upon an entire scientific misapprehension, and with 180 spaces between its two fixed points instead of the easily workable 100 . . .

Many people, even those who had been to a 'good' school, would have received little in the way of formal science teaching almost until the beginning of the present century. However, in 1871 a Mr J C Morris submitted a novel plan for consideration by the London School Board to rectify the situation. His intention was to bring science within the reach even of those who attended 'ordinary' schools but it would certainly have required a great deal of stamina on the part of the teachers who would put it into effect. What is more the pupils would have had to pay extra to attend science lessons.

The following are the principle points of the proposed system:—Subjects—chemistry, heat, light, sound, electricity, magnetism, telegraphy, mechanics, hydrostatics, steam engine, &c.; geology, mineralogy, metallurgy, botany, zoology, animal physiology, health, &c. A committee should be formed to select, revise and compile a set of suitable text-books, which should bear their sanction, and then be published in the cheapest possible form. There should be a depot to provide apparatus at a cheap rate, a complete set of which, sufficient to illustrate the sciences mentioned would not cost more than £100, and should be divided into ten cases of £10 each, a case to be completed for one or two subjects. The teacher *should be a visiting one* [my italics]. He could attend from two to three schools per day, and give from one to two hours' instruction in each, during two days in the week. The instruction to be given in a separate department, if there be one; or, if not, at such a time as would not interfere with ordinary school business. A single teacher could thus attend from six to nine schools weekly, if sufficiently near each other, and get through at least three or four subjects annually, so that in two or three years he would have completed the full course in each school. There should be an institution where teachers would have an opportunity of acquiring practical knowledge of their profession, and affording a means of testing their qualifications. A more economical way, however, would be for each teacher to have an assistant, by which method a nucleus of teachers would soon multiply into a goodly number. In conclusion Mr Morris advocates periodical examinations, with a regular system of rewards; and, in reference to funds, he thinks that the teachers and inspectors might be supported either by Government or subscription, the apparatus to be supplied by Government at reduced rates. Schools could fix a small fee for attending the class, which would add to its importance and help to defray expenses. Examination fees in like manner. Evening classes for adults could be managed under somewhat similar conditions.

This then is the background. What follows must be set within that context. In making a selection of items I have grouped them into very broad categories—science in the workplace, in the home, and so on. Whether all the items belong in the allocated groupings, or even whether they *are* strictly *science*, I leave for others to judge. Certainly a hundred years or so ago 'science' in the public mind embraced a great many things. For me it is significant that in the passage quoted earlier, Science and Art go hand in hand together. That is how it ought to be.

It must be said at the outset, however, that although this book might be regarded as filling a gap in social history, it is *not* a serious analysis of the impact of science in the nineteenth century. To have attempted a scholarly study of that kind would have been entirely inappropriate. For me the various items speak eloquently enough for themselves and very little explanation or commentary is required, although it is necessary on occasion to read between the lines. This anthology is intended to be no more than an affectionate look at what our forebears were thinking about and doing during the age when Queen Victoria occupied the British throne.

Chapter 1
Practical Science and the Useful Arts

IN Victorian Britain, well educated people of 'good' social background usually started work in an office job of some kind—often in some branch of the Civil Service, or possibly occupying a junior managerial position in the family business. The 'labouring classes' on the other hand had a much tougher start in life. Those right at the bottom of the social scale might even be forced to start work to help the family income at the tender age of six or seven. Those from that kind of background who succeeded as entrepreneurs later in life must have been truly remarkable people. Of course very very few turned their 'rags-to-riches' dreams into reality—not everyone had the necessary persistence or strength of character to keep on making sacrifices to achieve their long-term goal, and not all that did were lucky enough to have the necessary percentage of good fortune in the shape of opportunities at the right time. Those who did succeed in the end probably paid for it by becoming hard to live with and were likely to have been oppressive parents, demanding sacrifices from their children comparable to their own.

One person who began life as the son of an impoverished factory worker in Lancashire was a William Hoyle. When he died in 1886 he had become a wealthy mill owner entirely by his own prodigious efforts. In his obituary in the *Bury Times* (27 February 1886) it was reported that he was so determined to improve himself that as a boy and young man he attended night school after completing his normal day's work. In order to keep up with his studies, he got up each day at three o'clock in the morning.

Education, of course, was the key to advancement. Anyone who was unable to read or write would have been faced with an almost insuperable obstacle in running any kind of business for themselves. An interesting article in the *English Mechanic* in 1896 showed that, even then, it was realised that it was important for education to be *relevant*. The argument behind the article is surprisingly modern in the sense of advocating closer effective links between education and industry.

What is the use of teaching our youth the use of tools, unless, at the same time, we show them the application of what they learn to the vast manufacturing industries, at once our country's mainstay as well as its glory and pride?

When we teach drawing, should we not bring them face to face with the working drawings as used by the makers of our machinery, the weavers of our lace and other fabrics, the builders of our bridges, the makers of everything from the giant locomotive down to the penny toy?

When we put tools into their hands, what additional interest we might cause them to throw into the use of those tools could we place them side by side with the gigantic machine tools of our mills, our engineering shops, and let them compare their puny efforts with the irresistible force and absolute accuracy of the steam-driven implements.

Acting on these principles, the senior students of the metalwork classes held under the joint committees of the City Guilds and the London School Board, at the Marylebone High Schools, paid a visit, under the guidance of their instructor, Mr Wilson, to Messrs. Clayton and Co.'s engineering works, Westbourne Park, on Thursday last.

After examining the two boilers, and noting the method of firing, with the application of the steam-blast to liven up the glowing embers, they inspected the motive power of the whole shop. This was a two-cylinder engine working up to 60 HP. Close by, an engine for compressing air stood, and this was, after explanation, set in motion, and worked a rock drill, which in three minutes drilled a hole 8 inches deep and $1\frac{1}{2}$ inches diameter in a mass of hard stone. ...

For any young man who had scrimped and saved every penny he could with the intention of setting up his own business, it was a problem to know what would be a safe, yet promising area of activity in which to invest his hard earned capital. As in so many spheres, the answer was forthcoming in the pages of the *English Mechanic* as this 'Hint to Working Men in Want of a Living' dating from 1892 shows.

In these days, if a working man is the possessor of capital to the extent of £200 or £400, he often is at a loss to know how he should lay out his money. I believe there are openings in the outskirts of some large towns for men to collect and crush or grind bones for farmers' or market gardeners' use. The capital required is not large; but the work must not be commenced on too large a scale. First, having decided on a suitable locality where there seems the prospect of buying the bones from the kitchens of hotels or private houses at a reasonable price, the would-be bone crusher must either collect them himself or buy off a rag-and-bone man. If he collects them himself, he had better become at once a teetotaler, as often the bones may be had for a quart of beer at the public house round the corner, and the purchaser will naturally be expected to take his glass also.

After the bones are collected, they require boiling to extract the fat, which is sold to the soap makers. The boiling must be done in some out-of-the-way place as the smell is very disagreeable.

The would-be bone crusher will require the following plant, as it will probably take (at first) some months to collect a stock of bones on which to commence operations. The machinery can be quietly selected from the large quantity of second-hand machinery now in the market. First, a bone-mill; these when new cost from £30 to £100 or more—doubtless a very useful one could be got for £20 or £25. Having bought a mill, the next thing must be the engine. The small mills take 3 HP or 4 HP engines, the larger mills 10 HP to 15 HP. As the mill must not be too large, a 4 HP or 5 HP engine would answer. With a careful enquiry and inspection, such an engine might not cost more than £50 or £60.

Possibly a disintegrator may be afterwards bought, a shed for the mill, a tarpaulin for the engine, and a driving belt or two complete the 'crushing plant'.

Of course it is absolutely necessary that the bone-miller should have some knowledge of machinery; but he must be a good salesman—it is useless crushing the bones if he cannot sell them. There is no manure so beneficial to grass-land as bones. Our market-gardeners would find it to their advantage to use them more freely. Bones dissolved in one-fourth their weight of sulphuric acid make superphosphate, a manure valuable on account of its quick action.

N Y THOMAS

Then, as now, there were always those sharp-eyed opportunists who could see a chance to get rich quick by deception or subterfuge. Above all, the guiding principle was 'buy cheap—sell dear'. An article in the *Ethical World* bemoaning, in 1898, that 'nothing lasts nowadays like it did years ago' uncovered various dodges in the weaving trade that had become prevalent and showed how the unsuspecting purchaser of cloth was actually buying a large proportion of additives.

In the earlier days of industry cotton warp was treated with a mixture of fermented flour and tallow which gave it tenacity and lessened the friction of the weaving process. When the Factory Acts began to drive the new capitalists to wider and more daring expedients than the exploitation of their hands, this harmless detail of the manufacture afforded opportunities formerly undreamed of for misplaced ingenuity. The Crimean war having sent up the price of tallow, it was found that china clay made a cheap and showy substitute; and, after a few protests from an honest minority, its use became general. During the cotton famine consequent upon the American War, a short-stapled yarn came into use which required more and more size; and this suggested a second stage in the evolution of this particular fraud. The proportion of clay to cotton had now reached the maximum which could be carried. All sorts of odd compounds had been tried in addition, even poisonous salts of lead, mercury and arsenic. Then the happy idea was born of adding deliquescent salts, so that the weight of the water might be added to that of chemicals and clay in the facing. 'Ball-sizing' enabled 150 percent of foreign matter to be added to the weight of the pure fabric; then the more adaptable 'cylinder process' was invented. Finally 'steaming' was introduced; and today the cleverest imaginable arrangements are in operation for the injection of steam into the weaving sheds. Some pretence is made of protecting the health of the operatives against this twin evil of dust and moisture; but the vital statistics of Lancashire, Yorkshire, and Cheshire tell their plain tale. The makers of yarn reach the same end by cruder means. In one of the largest mills in Lancashire I was taken by a thoughtless youngster into a great cellar, on the brick floor of which, an inch deep in water, lay numbers of baskets full of cotton 'cops' or bobbins. Careful calculation was made of how much water they would absorb, how much would evaporate before they reached their destination, and how much they could retain without appearing too palpably moist. Fraud on this scale cannot be quite concealed. It must be known and, as elsewhere, winked at in the trade. It is only the ultimate victim, whose short-lived shirts and gowns raise memories of the 'good old times' who never learns why cheap things are so very dear, and high wages fail to raise his standard of bodily comfort.

While it is undeniable that this kind of deception was practised, it does seem likely that the writer of the article was overstating the case in rather emotive terms. It is not unlikely that some of these 'dodges' were necessary in new industrial processes, rather than as crude frauds. It is only to be expected that someone whose job might have been lost because of new technological developments would complain of unfair practice. What is surprising, to this writer at least, is to find to what extent new developments in science and technology were putting people out of work in the nineteenth century. We tend to think of that as a peculiarly microchip-based phenomenon of the late twentieth century. This article, gleaned from the *Acadian Recorder* (Halifax, Nova Scotia) of 19 February 1868, shows what the situation was like in the New World.

In Montreal large numbers of people are out of employment and there is consequently much destitution and distress prevailing. In Quebec between ten and fifteen thousand of the population are suffering for want of the necessaries of life, and deaths from starvation have already occurred in that city. The same state of affairs prevails in the United States. On a recent Saturday night one station-house in New York accommodated ninety lodgers, men, women and children, all persons who were guilty of no crime, but whom want and exposure had driven to this their only chance to keep from famishing. From North and South alike comes the tale of suffering. The destitution is so universal that it may very naturally be inquired: What is the cause of it all? The crops last year throughout the world were on the whole good, and there was not in 1867 any great commercial crisis to spread bankruptcy over the country. We can only think of one solution to the question. Whence this scarcity of work? The world is as large as ever it was, and the commercial world indeed is constantly being enlarged. We are afraid that Machinery is King. Man has become too ingenious for his own day and generation. Where, ten or twenty years ago, one hundred men were required to perform certain work, one man can do an equal amount in the same time and with greater precision and nicety. Steam and skill have done wonders within that period in revolutionising the world of labour, and the inanimate seems to be in a very fair way of superseding and crowding out of existence the animate workman. Where wharves would be crowded with men in days gone by to assist in loading and unloading vessels, steam appliances now perform the work in a tithe of the time, and at a far less expense. The mechanical world is commencing to find its occupation gone. The workshop is now a complication of whirling wheels, boilers, and belts. Skill and ability is greater now than mere animal strength, and the possessors of only the latter, as there are thousands among the uneducated classes, find that the field for their labour is becoming wonderfully abridged. So when hard times come, or there is a slight fall for even a short period in the business of the world, the cry of destitution is raised from many quarters at once, as in the present cases.

All we need to do today is to substitute microchips in place of steam in that article and it becomes immediately up to date almost to the letter, except that, fortunately a more caring society and the provision of financial help for the unemployed has kept at bay the stark deprivation and starvation of those former times.

Another point of great interest to me is that despite the advent of steam and all that it could do to allow jobs to be done more quickly, more cheaply, and with less labour than before, tremendous efforts were made to find still cheaper and more efficient alternative power sources. I K Brunel, for example used a water balance to lift building materials for his various bridges and railway stations when they were under construction. In 1865 a similar system was used in Paris by a M Léon Edoux, where it was claimed the water pressure was enough to lift it higher than the tallest buildings. An interestingly unusual type of water wheel, also of French origin, was in use in 1897 for irrigation purposes at Chateau La Planche, Vienne.

This device—called by its builders (De Coursac and Pascaul) a 'Vivonnaise' was used to lift water from a stream to a higher level so that it could flow into irrigation channels. It was claimed that with a current of 60 centimetres per second in the main stream, the wheel could raise 24 000 litres an hour to a height of 1 metre. The floats or paddles were made of corrugated iron, and each carried a zinc pipe curved in such a way as to lift up the water which rushed into it as the wheel turned. The whole thing could be permanently mounted between two columns or it could be rotated between two anchored boats which carried suitable bearings for the wheel shaft.

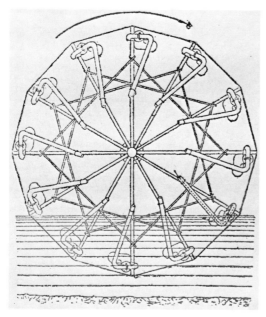

The Vivonnaise water wheel, 1897

Mr R W Hill, who will be mentioned again in Chapter 3 in connection with his invention of an audiovisual learning aid intended to teach parrots to talk, also had some thoughts about solving Britain's energy problems cleanly and economically once and for all. All that was required was the invasion and takeover of Iceland so that Britain could tap the geothermal energy under Iceland's surface, convert it into electricity and convey it to Britain by means of an 800 mile long power cable under the sea. Apart from the political questions associated with such a move, technically the idea doesn't now seem quite so far fetched as it must have done in 1894 when it was put forward.

The only technical problem with Mr Hill's proposal is the conversion device—what he called a 'steam battery'—for producing the electricity in the first place. This is what he had to say about it:

There are many persons who try to get electricity direct from coal. It is said that the so-called 'wizard' (T A Edison) is convinced that he can do it. I may, however, point out that what is possible is not always economical.

Moreover I would not be surprised if Prof. Dewar by subjecting the carbon (or a mixture of same with other elements) to an intense cold would produce an electric current, for the simple reason that his recent experiments showed plainly the fact that certain solids and gases subjected to an intense cold develop or conduct highly electric and magnetic properties; therefore, if he puts some carbon, say, with two platinum wires in his refrigerator for the liquefaction of atmospheric air, he would probably get electric current of high frequency from such an 'arctic cell'. Moreover, T A Edison is but a human being, and *humanum est errare!*

My opinion, however, is that instead of trying to get electricity direct from coal, it would be much wiser to try to get electricity from steam, and then of course the present nonsensical pretty-looking dynamo will be doomed. In fact I think that in ten or fifteen years hence the present dynamo would be regarded as a great nuisance, nonsense, and curiosity.

You will see presently that I have a substantial reason to say so, as I may point out now that we can get magnetism and sparks from steam, therefore there must be a certain contrivance by means of which we could rig up a cheap and efficacious 'steam battery'—i.e. a battery which generates strong current by passing of steam (or hot water) through it. Just think how envious would be our American cousins, who boast so much with a quite unnecessary gusto of their 'Niagara in harness' if an enterprising John Bull puts his steam batteries in all geysers of Iceland, connects the said isle to the British Isles by means of a cable 800 miles long . . .

I should like to draw your attention to an invention of a Swiss gentleman. The original notice appeared in *Electrotechnische Zeitschrift* for 1890. The apparatus consists of a hollow zinc ball 50 cm in diameter; inside of it a solid copper ball 40 cm is placed. If now steam passes between them, and at the same time they are rotated at a high speed in opposite directions an enormous pressure of current is generated. This electricity, however, is soon swallowed (so to speak) by the heat produced by steam, friction, &c. That, of course, is a great drawback of this invention. It shows that a certain combination of rotating zinc, copper and circulation of steam generates an enormous quantity of electricity; therefore it is very probable that we may yet find out this 'steam battery'. I know also a primary battery which generates dynamic electricity by passing off steam or hot water through same. Just think how such a battery would revolutionise the present nonsensical production of electricity. Why, every building which is provided with a hot water system would be also provided with steam, or rather, hot water batteries. You are warmed and your home is lighted at the same time through only one medium of source—i.e. hot water.

I admit I am a bit of an optimistic mind, though, considering the great progress made every year, month and week in science and industry, the dream of the year 2000 AD may be fulfilled many decades before the end of the twentieth century.

Lest anyone jumps immediately to the wrong conclusions, it might be wise to consider the slight misunderstanding Mr Hill himself displayed. He seems to have confused 'enormous quantity' of electricity with 'enormous voltage', i.e. a high *electric potential* of possibly several thousand volts—but with no substantial continuous electric current which is what would be needed for a power source. What was being discussed was really a variation of the 'Kelvin water dropper' device for producing *static* electric charge. No doubt one of the considerations in Mr Hill's mind was the long-standing prediction of doom due to the increased combustion of coal, and the consequent generation of excessive amounts of carbon dioxide (not to mention acid rain). Indeed, it had been predicted that civilisation would come to an end in about 1950 because by then the atmosphere would be too poisonous to breathe.

The phenomenon of obtaining an electric charge from steam directly had been known as a strange curiosity since Richard Trevithick's experiments with high pressure steam engines right at the beginning of the nineteenth century. Some unfortunate person accidentally received an electric shock from coming too close to a leaky boiler. This phenomenon later came to the attention of Michael Faraday but no serious efforts were made to harness the electricity produced in that way. In 1874 the *English Mechanic* carried a description of an apparatus designed by and made for Sir William Armstrong of Newcastle in order to conduct experiments in producing what was called hydroelectricity. The description is as follows:

A boiler about 6 feet long, 3 feet in diameter, fitted with pressure gauge, water gauge, and safety valve loaded to about 80 pounds on the square inch, rests on four strong insulating glass legs secured to a base plate; four (or more) steam pipes lying side by side proceed from the steam-chest, each pipe being fitted with a nozzle of box-wood; a conductor of stout brass wire carrying a frame with several metal points, is insulated and attached to the base plate, and capable of adjustment in front of the jets; a chain attached to the rod reaches the ground. The essentials requisite for obtaining powerful streams of electricity are, high-pressure steam issuing from small nozzles of either box-wood or ivory, perfect insulation of boiler, and the conducting-rod in connection with the earth (by throwing the *boiler* in connection with earth, and insulating the *conductor* an opposite electricity may be obtained.) The friction of the particles of water issuing violently through the apertures is supposed to be the cause of the electricity so powerfully developed. The circumstances under which the phenomena were first discovered is well known. I have been at a colliery where the men were so much startled at seeing sparks issuing from the steam as it escaped from the safety valve upon their thrusting an iron into it; but when the conditions above described are attended to, far more startling effects may be induced from a small generator of steam.

Kelvin's explanation of the effect is that the water droplets are ionised as they break up and the result is that the boiler and the spiked collector become oppositely charged and in due course sparking occurs in the gap between the spikes and the jets. The energy for this to happen, of course, comes from the kinetic energy of the steam issuing at very high speed from the nozzles. Although the phenomenon seems to have no practical value, who knows—it *might* just be the basis of space travel in the distant future. Odder things have happened.

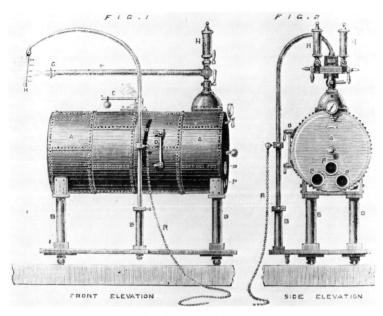

Hydroelectric machine, 1874

A much more practical form of hydroelectricity—using the term in the modern sense—was put forward in 1900 and had many similarities with proposals discussed in recent years in connection with the idea of a tidal barrage across the estuary of the River Severn. It was proposed to construct a funnel-shaped channel at a convenient location near the waterfront close to New York—a shape made naturally in the Severn Estuary. The idea was that as the water entered the channel under tidal influence, it would be concentrated into a progressively narrower section until it reached a point where the energy could be applied to rotate a turbine linked to an electricity generator. It was estimated that a channel with an opening of only 50 feet could yield a useful power of 100 horsepower. A more ambitious (and proportionately less practicable) idea was put forward by a Mr Elwyn Ryan. His device was called a water motor, consisting of a rocking float anchored some distance off shore. The energy was to be transmitted by long sprocket chains to a generator on shore. He estimated that with a float 36 feet in diameter, some 1500 horsepower would be generated. He seems to have conveniently overlooked the problems he would have had with his chains. A similar scheme by a Mr Ed P Butts, of Hartford, Connecticut had cogwheels instead of chains, but the business end of the device was to be four 30 feet diameter under-shot water wheels having paddles 40 feet wide. A really ingenious idea, however, was to use the rolling of ships to drive their own propellers—what a marvellous way to cross the Atlantic for absolutely nothing.

A liquid of a different kind that was thought to open up tremendous power generation possibilities—including transport propulsion—at next to no cost was liquid air. This report appeared in the *English Mechanic* on 24 March 1899.

A STUPENDOUS INDUSTRIAL REVOLUTION AT HAND

We are informed of 'a new substance that promises to do the work of coal and ice and gunpowder.' The new substance is, after all, a very old friend—nothing less than the common air, only in a liquefied state. Liquid air is the new mechanical magician. It was only in 1877 that Raoul Pictet succeeded by combining intense pressure and intense cold in liquefying oxygen. Fifteen years later Olzewski liquefied nitrogen and James Dewar actually solidified air—produced 'air ice.' The development of these discoveries into the production of a new and potent industrial force, as described in this paper, is the work of Charles E Tripler of New York City.

The principle is intelligible and simple to the last degree. It lies in the immense expansion in volume which takes place in any substance on passing from the state of liquid into the state of gas. When water passes into the gaseous state as steam, we have the force which drives our steam machinery. When liquid air passes into the gaseous state we have the new force. The immense difference appears in the fact that to change water into steam we have to use costly artificial means to raise the temperature above 212°F, while the transition from liquid to gaseous air takes place at 312° below zero Fahr. Once we have our liquid air, the temperature of the ordinary atmosphere raises it more than 300 degrees above its ordinary boiling point. In other words, the heat of the sun, in warming our atmosphere so much above the boiling point of liquid air is the ultimate source of the new power. Mr Tripler claims to produce liquid air 'practically without cost.'

A single cubic foot of liquid air contains 800 c.ft. of air at ordinary pressure—a whole hall bedroom full reduced to the size of a large pail. Its desire to expand, therefore, is something quite irrepressible.

Actually how the liquid air is produced in 'commercial' quantities, Mr Tripler neglects to say, except that somehow he seems to be putting a great deal of strain on the second law of thermodynamics.

The liquefaction of the air, according to Mr Tripler, 'is caused by intense cold, not by compression, although compression is a part of the process. After once having produced this cold I do not need so much pressure on the air which I am forcing into the liquefying machine. My liquefying machine will keep on producing as much liquid air as ever, while it takes very much less liquid air to keep the compressor engine going. This difference I save . . . . I have actually made about ten gallons of liquid air in my liquefier by the use of about three gallons in my engine. There is therefore a surplusage of seven gallons that has cost me nothing, and which I can use elsewhere as power.' It would indeed be wonderful if he could. What is more he claimed to be able to produce about 50 gallons a day at a cost of no more than twenty cents a gallon.

The rest of the report goes on to list some of the amazing things that could be done using liquid air.

A few drops retained on a man's hand will scar the flesh like a white-hot iron; and yet it does not burn—it merely kills. For this reason it is admirably adapted to surgical uses where cauterisation is necessary; it will eat out diseased flesh much more quickly and safely than caustic potash or nitric acid, and it can be controlled absolutely. Mr Tripler has actually furnished a well-known New York physician with enough to sear out a cancer and entirely cure a difficult case. And it is cheaper than any cauterising chemical in use . . . It freezes pure alcohol . . . Mercury is frozen until it is as hard as granite . . . Iron and steel become as brittle as glass . . . . The experiments have been shown by Mr Tripler before a meeting of distinguished scientists at the University of the City of New York. Among the number was M. Pictet, who expressed great admiration.

But the prospect opened up of future possibilities is dazzling and bewildering in its grandeur. Think of the ocean greyhound unencumbered with coal-bunkers and sweltering boilers and smoke-stacks, making her power as she sails from the free sea-air around her. Think of the boilerless locomotive running without a firebox or fireman, or without need of water-tanks or coal-shutes, gathering from the air as it passes the power which turns its driving wheels. With costless power, think how travel and freight rates must fall, bringing bread and meat more cheaply to our tables and cheaply manufactured clothing more cheaply to our backs. Think of the possibilities of aerial navigation with power which requires no heavy machinery, no storage batteries, no coal . . . .

Of course it is one thing to give a small-scale demonstration in a university lecture theatre, and quite another to turn it into large-scale industrial reality, no matter how attractive it might be as a speculative idea. Mr Tripler conveniently concealed the energy that had to be expended to produce liquid air in the first place, and which must add to the overall cost. What is more, all these liquid-air-powered ships, locomotives and aeroplanes would have to carry with them the machinery to liquefy the air that is driving their propulsion engines, as well as the necessary fuel. In the end they would be carrying more than if they used the original fuel in the first place in the conventional way. What happens too when the liquid air turns the steel containers into the equivalent of fragile glass?

Nevertheless the idea of using liquid air, liquid carbon dioxide or liquid hydrogen as a power source as yet has come to nothing. Still there is something that continues to attract fertile minds because in principle a gas expanding as it warms is no different from what we already do, albeit within a different temperature range, with our present prime movers. If the liquid air or other substance happened to be produced as an unavoidable by-product of some other essential industrial process, then we can ignore the cost of production in the equation. We would then simply be using fruitfully what would otherwise be thrown away and wasted. The lessons of history, however, suggest that we would still have to pay for this 'free' energy in the end. Petrol was once thrown away because there was no use for it.

Before 1860, when Etienne Lenoir made the first practical internal combustion engine, there was no significant use for petrol other than for medicinal or dry cleaning purposes (the term 'petrol' was actually introduced as a trade name many years later and was not in use in the 1860s). There was, however, a petrochemical industry of a limited kind mainly geared towards producing lamp oil for illumination and furnace oil for heating purposes. The following article describes the process used in 1865 for refining crude oil.

In the treatment of petroleum almost every part of it is brought into merchantile condition, even the residuum, or coke, which remains after the various oil and gases have been obtained is turned to account for fuel, being used to carry on the several chemical processes to which the crude petroleum is submitted. The products obtained are—benzine or naphtha, burning oil, paraffin and the coke. The naphtha, which is the lighter element, being something between a gas and a fluid, is driven off from 90° to 300° of heat, and constitutes about 15 percent in bulk of the crude oil. The lighter naphtha is now called gasoline and from it an illuminating gas is prepared. The warmth of the hand will cause it to boil. A heavier distillation is used for cleaning purposes, and a still heavier one has been used by painters as a substitute for spirit of turpentine. At from 300° to 500° the product of the still is a light oil, which, after treating with sulphuric acid and caustic soda, for the purpose of deodorising it, becomes the burning oil of commerce. This constitutes about 60 percent of the crude oil. As the heat is gradually increased from 500° to 800° heavy oil is produced, which, being treated in the same way as the burning oil, is then barrelled up, and packed away in the cellar, with ice and salt, for the purpose of chilling it, as it is called. This is effected shortly, when, upon opening the barrels, the oil is found to have assumed something of the appearance of calves-foot jelly. It is then put into canvas bags, and being placed under a screw press, is subjected to an immense pressure. The finished lubricating oil is thus squeezed out, and separated from the paraffin, which is found in the bags in the shape of thin white cakes, which are almost perfectly tasteless, and in every respect resembling wax. This being purified is used for the purpose to which common wax is applied, such as the making of wax candles, clear starching, the manufacture of chewing gum &c. In its refined state this paraffin is worth from 20c. to 35c. per pound. The lubricating oil will perhaps make 9 to 10 percent of the amount of crude oil from which it is distilled. The paraffin wax may equal 1 percent. At 800° the contents of the still will show a red heat. The distillation is then stopped and the residuum, which is a sort of coke, similar in appearance to ordinary coal coke, is removed and used for fuel.

An unusual use for a gas of a different kind was reported in 1897 in *Le Monde Moderne*.

Acetylene, already so much spoken and written of as an illuminant, threatens to introduce itself in the confection of liquors. This use of it may, at first, appear but little appetising, when its disagreeable odour is considered; but we hasten to say that care is taken to transform it into alcohol, for it is a gas which contains the principal elements of that precious liquid; there remains but to add what is lacking—oxygen.

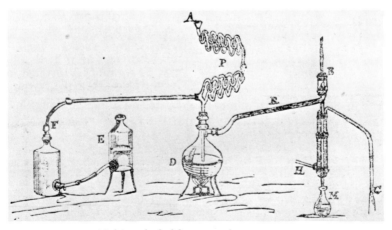

Making alcohol from acetylene gas, 1897

Alcohol is, in effect, a compound of carbon, hydrogen and oxygen; acetylene possesses already the first two elements. We will now give it the third and increase to completion the dose of hydrogen, which was a trifle too small in quantity to bring acetylene quite up to the alcohol stage.

In a flask F, calcium carbide and metallic zinc are placed; knowing that zinc, when attacked by water acidulated with sulphuric acid, gives hydrogen gas in the presence of water, we see here that the calcium carbide freely evolves acetylene gas. In the flask E put water and a little sulphuric acid, and connect this flask with the first by a flexible tube so that, when E is elevated or lowered, we may introduce or withdraw at will, liquid in the flask F according to the need of the production.

The acetylene and the hydrogen, developing at the same time, do not fail to seize the opportunity for combining. In the nascent state bodies always have a greater affinity for each other than at a later stage.

It is of this marriage, then, that the ethylene is born, which, being now disengaged, goes over into the glass worm P where it comes into contact with concentrated sulphuric acid heated to 80° Centigrade, which is slowly poured into a funnel A; it is here that it gets its oxygen. It now forms a new body which is ethyl-sulphuric acid. This is collected in the flask D and is brought to ebullition. Here it is decomposed into sulphuric acid, which remains and may be used again, and into alcohol which evaporates, but is collected and condensed by means of tube R connecting with worm B surrounded by a current of cold water circulating from H to C.

In M is collected an alcohol absolutely pure, which industrially produced would not cost more than 20 centimes (4 cents) per litre. It contains none of those essences which are always present in the vegetable alcohol, and which render them dangerous for consumption.

It is not a little curious that alcohol, furnished in large quantity by the vegetable kingdom, is now going to be given us by the mineral world, and at a ridiculously low price. It would seem that in combating alcoholism we are going against the laws of nature. After all, we had best conclude that if Dame Nature thus places alcohol in profusion within our reach, it is not to take the place of water as a beverage.

A possible use for alcohol made in this way might be for 'medicinal purposes'. An article in the *Medical Press* in 1896 drew attention to the use of beer to treat ailments of cattle and horses and highlighted this as what must be a highly original reason for the renewal of a licence to sell alcoholic beverages.

The use of alcoholic beverages as a medicine for cattle and horses is common enough in some parts of the kingdom. A short while ago an application was made before a Welsh licensing session for the renewal of an off-license for a place bearing the somewhat outlandish name of Bryn-y-Bual. The applicant stated that the house in question was situated in an agricultural district, and was a great convenience for the farmers to get beer for their cattle in times of sickness. Speaking generally, there is no apparent reason why alcohol should not be just as valuable in the treatment of certain morbid conditions in cattle as it undoubtedly is in the higher animal, man. The principle of stimulation as an active therapeutic agent applies with equal cogency to both classes of mammalia. It is one, however, that is likely to raise a regular whirlwind of commotion among the teetotalers, who, as a body, are not given to attach much importance to purely scientific considerations. A new field might open itself to that excellent class of reformers in the discussion of such an abstruse point as 'Should cows drink Cognac?'

Air in its natural state has long been a source of power. A Mr W F Martin in 1869 came up with a simple, yet quite effective, idea for making use of the movement of trees rocking in the wind. The illustration shows how the idea could be applied to operate pumps—either directly, or by an indirect system of bell-cranks and levers. The latter looks rather a risky arrangement in a storm—in the event of the tree moving violently in more than one direction there seems to be a grave risk of pulling off the corner of the building. For some strange reason, Mr Martin found it necessary in his description of the device to point out that it was not suitable for propelling vehicles.

Martin's wind-powered pump, 1869

A more conventional 'wind motor' was devised in 1894 by a Mr H P Saunderson of Kempston Road, Bedford. He described it as 'the most mechanical and scientific device yet invented for furnishing cheap motive-power' which says something about his ego.

The machine was rated at 'two horse-power in a fair wind' and was sold in knock-down form ready for erection on the farm. Mr

Saunderson's supreme self-confidence stands out in his description of the motor and how it was operated.

Saunderson's wind motor, 1894

Compare the illustration with that of any other make. You see in this a trim, symmetrical and well-balanced mill, as durable as iron and steel can make it, and capable of doing work in a wind so light that it would not cause an ordinary windmill to revolve; yet it is under perfect control in the strongest winds. It is self-regulating and automatic in action; as soon as the starting handle is pulled, the motor immediately faces the wind and starts. In case of violent storms, the wind wheel regulates off and presents its edge to the wind.

The main bracket containing the gearing is cast in one solid piece which cannot rack or become loose, and the vertical shaft is geared to run at such a speed that machinery can be driven from it direct, or by belt or gear without further speeding; this enables a much lighter shaft to be used. But the most important of all is that by the application of this principle of gearing, the twisting tendency of the vertical shaft to turn the motor out of wind is not only rendered harmless, but is utilised to bring the face of the wind-wheel up to the wind, increasing its power in proportion to its work, at the same time rejecting surplus power caused by gusts of wind and storms.

It is very easily erected. The whole machine is attached to a stout pitch-pine mast, which can be fixed in a barn or other building, or it may form the principal of an independent tower, as the case may require. When the mast is erected, the different parts of the motor can be bolted in their respective places, and within twenty-four hours it is ready for work.

It is particularly adapted for farm work, such as grinding corn, chaff cutting, pulping roots, pumping water, or for driving dynamos, lathes, grindstones &c., or for any other purpose where cheap motive power is required. It is unequalled for working duplex or double-barrelled pumps, and where such are already in use, the horizontal shaft of the motor is connected direct on to the existing plant. There are few farms in this country on which it would not save its entire cost during one winter in grinding and chaff cutting alone. When provided with a large hopper, it can be left grinding all night with perfect safety—in fact it is specially designed to run practically without attention; a little oil occasionally is all that is required.

Although the illustration conveniently obscures the vital point of the mechanism at the top of the drive shaft, it rather looks as though the tail fin is attached to a simple differential gear, but it is not possible to tell exactly how this shifts the wind-wheel edge-on to the wind as the description says it can. The horizontal shaft on which the wind-wheel itself revolves appears to have to substantial support and has a dangerously long overhang. However, the basic idea is sound enough.

An interesting possibility for the wind motor might be to provide free heating for factories. In December 1865, the *English Mechanic* in its 'Useful and Scientific Notes' suggested that 'Two iron plates 4 feet in diameter, and weighing 1600 pounds, revolved at 80 revolutions per minute send sufficient heat up a furnace to warm a large factory . . .' Presumably the idea was to rotate these discs in opposite directions, in face-to-face contact. While they would certainly get hot, they would also let out the most unimaginable screeching and grinding to an unbearable degree. Apart from noise and dangerous chemicals in the factories of the nineteenth century, another hazard was the machinery itself—either the operator getting trapped in unguarded cogwheels, or bits of machinery breaking or coming loose and flying off in all directions. Professor C H Benjamin, in *Cassier's Magazine* for July 1900 drew attention to narrow margins of safety, especially in flywheels on engines made to rotate at unsafe speeds.

As in the earlier boilers, working with steam at atmospheric pressure, explosions were unknown and safety-valves yet to be invented, so in the engines of the last century the fly-wheel had no thought of flight as it soberly made its ten or twelve revolutions per minute. The 19th century has changed all that. The horse trots in 2.04 instead of 2.40, the bicycle and the automobile go a mile a minute, the Atlantic 'greyhound' and the Pacific 'flyer' continue to cut records, while financial, social, and mechanical safety-valves are all blowing off at high pressure.

Within twenty-five years boiler pressures have crept up from 70 to 250 pounds, and piston speeds from 300 feet to 800 feet per minute, while improvements in strength of material and security of design have hardly kept pace; boilers and fly-wheels are exploding all round with lamentable regularity. It is of no use to cry 'halt'; the modern man will not halt; speed and pressure continue to increase, and measures must be taken to make the increase safe. The introductions of various forms of safety boilers for high pressures and the more stringent laws with regard to license and inspection of steam boilers will probably lessen the danger there. A fly-wheel is just as dangerous as a boiler, and should be subject to inspection in like manner.

The demand for a high rotative speed in engines used for electric power and lighting is responsible for many fly-wheel accidents, and it is also probable that many rolling-mill engines are now turning faster than their builders intended. The designs of many of the older wheels now in use are entirely wrong, especially in the feature of rim joints, and while the wheels are comparatively harmless at low speeds, they are now perilously near the bursting limit. Some years ago, the writer had occasion to investigate a large wheel of the rolling-mill type. He found the strength of the joints less than one-third that of the solid rim, and the factor of safety at the given speed about two. The old wheel has since ended its life, fortunately without killing anyone.

Both builder and owner are sometimes aware of the narrow margin of safety, and, when an accident happens, are anxious to prevent an investigation. The time to investigate a fly-wheel is during its lifetime, and the one to investigate it is a trained inspector who can pronounce intelligently on its safety or condemn it if dangerous.

A correspondent to the *English Mechanic* on the subject of boiler explosions in 1865, however, was at the receiving end of an explosion from the editor. He was told in no uncertain terms that much of his letter had been 'struck out' because he had ventured to criticise in too personal a way for the editor's liking, some observations on the subject by a previous correspondent. The writer, by the name of H Reynolds from Manchester, clearly didn't think much of theory, or those who could only speak from a theoretical standpoint. This is what the editor left of what he wrote:

A subscriber from the first I have been, and hope still to be, delighted with my weekly pennyworth—*The English Mechanic*. I am only a stoker, but I can understand a great deal of what is set before me in your paper, and by yourself and your correspondents, whom I have had to thank for information once or twice. There is one correspondent who signs himself 'F.M.' a new member of the 'Correspondence Club' who, in my opinion—if that is worth anything—had better leave off writing on boilers until he knows something about them. His writing is 'one of the things no fellow (in my line) understands,' and I don't think he's too sure about it himself.

I always notice that the less a man knows about a thing in which he would like to dabble, the more he goes 'supposing' and 'guessing' and making what he is pleased to call 'theory' do duty for 'practice'. I don't say 'F.M.' does this; but what he does is very like it. He says that the majority of explosions are shrouded in mystery. Well, if he writes of the appearances their particulars present to himself I am not qualified to doubt, but if he thinks to express my opinion in his declaration, I can tell him he is—well, he is utterly wrong. I am no better informed perhaps, than many other stokers, and yet I don't remember ever having heard of an explosion which could not be accounted for, if the parties 'interested' would only speak out. Now, down here, we have the Steam Boiler Association, and Mr Fletcher never experiences any difficulty in giving a 'why' or a 'wherefore' in every case submitted to him; and that being the case, it is evident to any thinking man that such explosions are preventable; consequently, that what they are shrouded in at the time of an explosion is not mystery by a long chalk.

More than that, it is shown by the officers of the Association, and by the boilers they look after, that there is no mystery in the matter, and that care and attention are the only things necessary to keep a boiler in long life.

If 'F.M.' will get hold of the reports of this Association he will find everything clearly accounted for; and he ought to know that a coroner's inquest is scarcely the place in which to find light thrown on a subject, constituted as some of these assemblies are. 'F.M.' has not asked for any information on the subject, but in case he should, I would ask him to canvass among the shops here for the opinions of stokers and engineers. I am sure he would think twice after that before he thought of asking you for more room, and to be put in big type.

I can't say I've studied the growth of steam over a fire; and perhaps 'Mr F.M.' is right in what he says about that. I have heard before that perhaps gas from the exterior could penetrate the boiler shell, and give fits to its contents; this I should imagine could be done by subjecting it to enormous pressure; but as this is not done in daily practice, I fancy that any gas which may penetrate to the outside of a boiler is rapidly driven away again, up aloft, or drawn by draught under the bars, through the fire and turned into the air as something else. But there, I'm almost getting as—as shrouded in mystery as 'F.M.' who takes longer in working to a point than did another 'F.M.'—'The Duke'; and therefore I will leave off, as my turn at stoking begins shortly. And I'm only repeating what my mate and the other 'shift' say, when I express to you the hope that the next correspondent who lays hold of a boiler for a subject will be careful what he is about, and not subject himself to a 'blow up', though, mayhap, he might see a 'flash of lightning' during the operation.

H. REYNOLDS

Not surprisingly, in the face of that counterblast, the unfortunate 'F.M.' chose to remain silent. The interesting thing I find about Mr Reynolds' letter is that despite being 'only a stoker' he had no hesitation in arguing in public with someone he clearly thought was 'talking through his hat'. It also reveals something of the suspicion about the value of 'theory' as opposed to practical experience that many working people felt in those days–and for many a year after.

Much of the correspondence with industry in the *English Mechanic* was on the level of 'practical hints' rather than to do with theoretical matters. the following are typical examples dating from 1865, 1899 and 1900 respectively:

HINT TO MECHANICS

Sir—In your invaluable work appear many articles which must be very useful and of great benefit to the working mechanic, and from which I for one have derived a great deal of information.

There is a trifling thing I wish to bring under the notice of joiners, pattern-makers, and all who use a whetstone with oil, that paraffin oil is the best for that purpose; it only requires a trail, but the stone must be free from all greasy oil before it is applied.

MACHANISTA

REMOVING RUST

Soak with petroleum—roughest dark stuff, if you can get it—for several days; it eats into the rust and can be scrubbed off with a wire brush. Brush over nearly saturated solution of chloride of tin, and rest 12, 14, or more hours; clean off; brush over with strong solution of cyanide of potassium—say half an ounce in a wineglass of water—then, with fibre or hair brush, with paste of cyanide of potassium, Castile soap, whitening and water to creamy consistency; proportions to be found. Sulphuric acid diluted 1, water 10; afterwards wash with hot lime, water, and dry off with sawdust if possible. Cover with sweet oil well rubbed in; 48 hours after rub with finely pulverised unslaked lime. Whichever of foregoing processes tried, and when fairly clean surface is got and enamel or paint is required, use a good maker's anti-rust; but if, in getting their list, you find they do not sell small quantities to suit you, ask for a retail customer's address most convenient where you can get supplied.

REGENT'S PARK

BURNT STEEL

Some time ago a querist asked if burnt steel could be brought back to its original state, but most of your correspondents seemed to think the only way was to re-melt it. In our shop (Mr Plenty's) we use some stuff which looks like a mixture of pitch and some kind of fats, no name and very little smell. It is kept in small buckets for the purpose. I melted the end off an old file, put it in the stuff (which soon melts, giving off clouds of vapour) and let it get cold. The steel does not need hammering; but after I had heated, hardened and tempered it, I broke it, and most certainly it seemed and looked as hard, fine-grained, and not to crumble any more than any other file. Since then I have used it for tools &c.

S.A.K.
Battenmere, Hungerford, Wilts.

I wonder what would happen today if someone walked into a chemist's shop and asked for half an ounce of potassium cyanide, giving the explanation that it was to remove rust from the bottom of the car. As for using paraffin (kerosene) instead of machine oil on a carborundum sharpening stone, that is an excellent idea. I have used it with first class results for sharpening DIY tools such as wood chisels and similar things.

Another example of 'practical' science hallowed by long tradition and with little or no reference to theory is cheese making. The following is an account of how Roquefort cheese is (was?) made. It was recorded in the *English Mechanic* in 1896.

It is supposed that hundreds of years ago the South of France was disturbed by volcanic eruptions which split up the ancient granite rock, causing streams of lava to flow from them. The new surface consists of basaltic rock, which in its turn was fissured by eruptions and thrown up on a mountain range. The whole of the interior of a mountain was formed into caverns and caves which belch forth hot sulphurous springs. It is here that the celebrated Roquefort cheeses are made. The village of Roquefort is situated on the Mountain Larzac which is about twenty-five miles in length and nearly 3,000 feet high. It consists chiefly of limestone covered with sufficient pasture to feed the 300,000 sheep kept for their milk. The caves, being formed by the displacement of rocks, consist of an intricate labyrinth of open spaces and passages connected with each other and with a subterranean outlet. A cool current of air, therefore, always of the same humidity and temperature, flows in a never interrupted stream through the caves. There is nothing in the milk or in the preparation of the cheeses that gives them that peculiar flavour and delicious mellowness for which they are so renowned. This is entirely effected by the method by which they are cured.

When the cheeses are ready for treatment they are taken to the caves, and after being allowed to cool are carried to the salting room. They are rubbed with salt on one face and then piled on the top of each other until the cave is full. After standing for twenty-four hours or so, the reversed side is salted, and once more they are piled up as before. The cheeses have to be frequently reversed in order that the moisture may be even throughout, and to develop the fungus which has previously been sown in the curd. In forty-eight hours the cheeses become viscous, and are rubbed with a coarse cloth. In the course of another two days the fungus will appear on the outside in the form of a sticky paste. This is carefully scraped off with knives, together with a thin stratum of crust, and set aside for food. The cheeses are now sorted out; the most solid ones placed on the floor. In eight days' time they become covered with a yellowish red mould, together with other minute vegetation, which is removed and given to the pigs. The scraping is continued until the character of the mould changes, showing that the curd has altered its condition, and announcing the completion of the cure. Then they are again carefully scraped and wiped, and wrapped in tinfoil, and are ready for the market. Roquefort cheeses have been cured for centuries by this process, and stand as a triumph of uneducated art.

Whilst on the subject of food and the food industry, an interesting report appeared in *Chemical News* in 1866 regarding the preservation of meat. The process was introduced at a meeting of the British Association for the Advancement of Science which took place that year in Nottingham.

Amongst the objects exhibited at the *soirées* and the Pharmaceutical Conference during the meeting of the British Association at Nottingham, few things attracted more attention than sundry amorphous-looking lumps, covered with a white coating. Labels told the visitors that one of these was a mutton chop, another a loin of mutton, a third a sirloin of beef, preserved by Redwood's process. This process consists in the immersion of fresh meat in melted paraffin, at a temperature of 240° Fahr. for a sufficient time to effect a concentration of the juices of the meat and the complete expulsion of the air; after which the meat, in its condensed state, is covered with an external coating of paraffin, by which air is excluded and decomposition prevented. The concentration of the juices may thus be carried to any required extent. If the meat is to be kept in hot climates, its weight should be reduced by evaporation to about one-half in which state it will contain all the nutriment of twice its weight of fresh meat, the portion driven off by evaporation consisting only of water. Thus prepared it will be fully cooked (by the heat applied in the process) and it may not only be eaten without further preparation, but it will also be applicable for the preparation of a variety of made dishes, including stews, hashes, soups, gravies &c. For cold climates a less amount of heating and concentration will suffice, so that the meat may retain its original juicy condition, and when further cooked, present the appearance and possess all the characters of fresh, unpreserved meat. The paraffin used in the process is perfectly innocuous, free from taste and smell, and is not subject to change from keeping. It may be removed from the surface of the meat by putting the latter into a vessel containing boiling water, when the paraffin as it melts will rise to the surface of the water, and may be taken off in a solid cake when cold. Among the advantages claimed for the process may be mentioned its great simplicity, the facility with which it can be performed by unskilled workmen, and its inexpensive character, as the same paraffin can be used for an indefinite number of times, and the quantity required for coating the meat is small. When the meat is concentrated as described, it is rendered very portable, and no special care is required in packing it. Samples of meat preserved by this process have been tested by Messrs. Gillon and Co. of Leith, the well-known preserved provision merchants, with perfectly satisfactory results . . .

The whole idea sounds utterly revolting—a process performed 'by unskilled workmen' and in which 'no special care' is required in packing it implies it might be dropped on the floor any number of times before it reached its destination.

Contrary to what the history books imply, the development of prime movers throughout the nineteenth century was not a straightforward 'linear' evolutionary process. The whole field was one of arguments about prior rights, dead-ends and blind alleys, and countless attempts to circumvent patent restrictions on orthodox machinery. One such was a brave attempt to construct a steam engine which did not have a reciprocating piston sliding in a cylinder. This was the work of a Mr Musgrove in 1869.

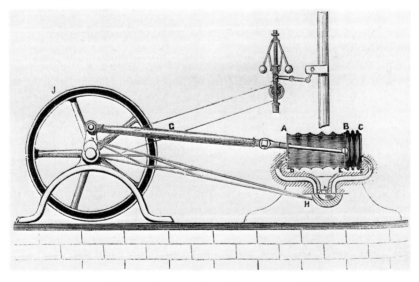

Musgrove's expanding cylinder engine, 1869

The idea was to combine the piston and cylinder into one expanding vessel, which operated concertina fashion under either steam or hydraulic pressure. Mr Musgrove first thought of using leather but found that it was too leaky, being porous. He then tried indiarubber reinforced with wire rings. The problem then was to stop the 'cylinder' bagging out sideways between the rings and bursting under the pressure needed to turn the crankshaft and flywheel—let alone do any useful work. Interestingly enough, although Mr Musgrove genuinely thought he had hit on a new idea, he had already been anticipated by more than 30 years. In the 1830s an engine of identical design was built and tried out by an engineer named Walter Hancock with a view to applying it to drive a steam-powered omnibus. He soon found it was impractical and abandoned it.

Another extremely unconventional engine—which did have some limited success in the 1860s—was designed by M Bourdon, the French engineer who invented the well known Bourdon tube. The Bourdon tube is the essential element in many mechanical pressure gauges. It is in effect a flattened tube bent into the arc of a circle and closed at one end. The other 'open' end is connected to whatever vessel contains the fluid whose pressure is to be measured. The action of the pressure is to tend to straighten the tube and, by means of a suitable linkage, the movement of the free, closed end is magnified and translated into the movement of a pointer over a calibrated scale. A sound engineering principle is, having found a good way of doing something, stick to it. This M Bourdon certainly did. His engine is nothing more than a jumbo-sized Bourdon tube—or rather two tubes joined together at their 'open' ends to a supply of steam or water through a suitable valve.

It is not clear from the illustration which appeared on the front cover of the *English Mechanic* for 4 September 1868 just how the working fluid was regulated and admitted to the tube. The principle of operation, however, is clearly the same as in the pressure gauge. As the tube alternately flexes and extends, it oscillates the almost vertical levers at each side. These in turn are connected to a two-throw crankshaft which has cranks at 180 degrees, by means of conventional connecting rods. These rotate the crankshaft in the ordinary way and the huge flywheel helps to even out the motion and to prevent the whole thing oscillating, once it has started. Apart from the obvious problem of metal fatigue eventually causing the tube to crack, a major drawback with the engine—especially if it is used to drive a lathe or similar machinery, is that it could just as easily start to rotate backwards as forwards, unless the engineman in charge pulled the flywheel round first. Moreover it was not possible to increase the power by increasing the pressure, except within a small range, because if the tube straightened out too much it wouldn't return to its normal state and would then stop the engine.

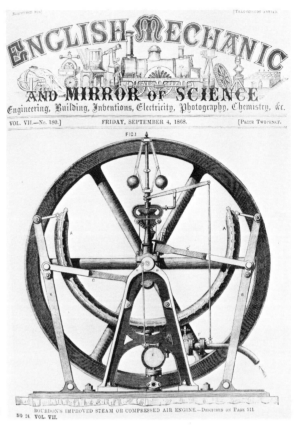

Bourdon engine, 1868

Another curious engine was invented and constructed by Mr B Wagenknecht, Constructing Engineer to Queensland Railways, Australia. Again it was an attempt to find an effective alternative to the reciprocating, cylinder and piston design and was one of the forerunners of the latter-day Wankel engine.

The interesting thing about this engine is that it could be run either as a steam-powered machine or as a paraffin-fired internal combustion engine. The illustration is fairly self-explanatory. A rotating blade carries with it an eccentric member inside a cylindrical chamber. At a certain point a valve opens and a jet of compressed air forces a fine spray of fuel into the engine where it is ignited by an electric spark. Expansion of the hot gas then carries the blade round until it uncovers a hole in the casing through which the products of combustion escape and the cycle recommences. It probably would have worked much better as a steam engine since, as an internal combustion machine, there was no provision for compressing the charge of fuel and air and that meant a significant wastage of potential power. With steam that would not matter. Another serious defect—and one shared by modern rotary engines—was leakage past the sides of the moving parts, especially after a slight amount of wear had taken place. It was a good idea while it lasted.

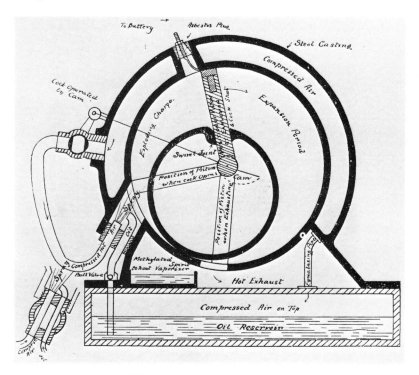

Wagenknecht rotary engine, 1897

Experiments in the use of hot air also attracted a great deal of attention since it offered what was effectively a steam engine without the problem of having a boiler. This design was actually intended for a motor car power unit although it never left the drawing board. It was the brainchild of an engineer named Alfred George Melhuish and was published in the *English Mechanic* in 1900.

... I have designed the above new form of engine that is based upon an old principle, the type being known as 'Stirling's Regenerative'. I have drawn the engine in section on a vertical plane in order that the whole interior action may be better understood. At first sight it would appear that the design simply embodied the combination of four distinct engines; but this is far from being the case because if any engine was removed, the remainder would not work. The engine is constructed upon what I term a four-cycle principle.

Melhuish hot-air engine

As will be seen, a crankshaft is employed having four cranks at an angle of 90 degrees to each other, each crank beign operated by its respective piston and connecting rod working in a cylinder placed immediately below, and in line with the crankshaft. Each motor piston is connected by a straight rod passing through a boss in the refrigerator plate to the displacer immediately below.

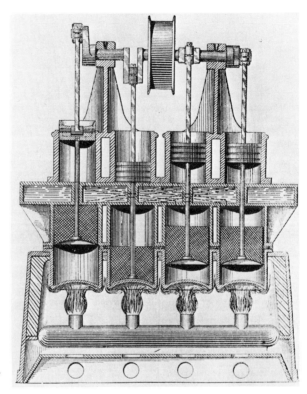

Melhuish hot-air engine, 1900

Now as most of our readers are aware, it is a peculiarity of hot-air engines of this type that the displacer should always move in advance of the motor piston a distance corresponding to the advance of one crank on the other of 90 degrees, and this has hitherto been done by employing either two cranks, set as above described, or levers have been so disposed as to accomplish the same result. All this has meant undesirable complication, and would not be suitable for the purpose we have in view. But in this case the displacer is connected directly to the piston, and each piston is successively operated by the displacer, the crank of which is in the necessary position for so doing. For instance, the first cylinder is operated by the third displacer, as the crank to which it is connected is 90 degrees in advance of the first crank. At the same time the motor piston of the third displacer is operated by the second displacer, inasmuch as its crank is 90 degrees in advance of the third crank, while the second piston is operated by the fourth displacer, and the fourth piston by the first displacer, thus completing the whole cycle. This engine, by reason of the equal disposal of its cranks and displacers, has no 'dead centres', is in 'perfect balance', has no need of any fly-wheel (a split pulley being shown in position on the crankshaft for power transmission), and for a given amount of thermal units will give 'greater power'; it has less frictional waste and gives an equal torque at any given position of the crank-circle and is therefore, I think, a distinct advance on anything of the kind hitherto seen.

Since making the drawing, I can see that the cranks would have been better placed in consecutive order, in which case the ports between motor cylinders and displacer cylinders would be as shown by the dotted lines. But this is a mere detail. In my next communication I will show how this engine can be made in circular form, without crankshaft, with only two connecting-rods and no motor cylinders and pistons (as such) so that the while thing can be put together by anyone possessing a drilling machine or even a fiddle-bow.

Mr Melhuish's idea does seem to have a great deal of merit and the idea of coupling up displacers and pistons in the way described is ingenious. To my knowledge, the 'circular' form of the engine never materialised, which is a great pity since it would have been interesting to see what he was proposing. For anyone with the skills and facilities it might be worth making a model of the four-cyliner in-line engine to see how it performs.

In addition to prime movers, a great deal of attention was also given to transmission systems using combinations of belts, chains and various kinds of gearing. One of the problems with nineteenth-century engines was the enormous bulk and weight for a limited power output. This was especially the case for powered vehicles, in the sense that a necessarily 'small' engine meant a feeble power output, and that in turn meant that in order to avoid stalling the engine when starting off under load, some way of gradually taking up the drive in as gentle a way as possible was needed.

Here is an example of such a system of gearing intended for use on a powered tramcar. It was designed by an electrical engineer in Birmingham named G F Chutter in 1894. It did not, however, meet with universal approval from fellow engineers, the reasons for which are explained later.

Let us see what are the chief requirements of a perfect gear. In the first place the drive must be certain. No friction drive can be admissible so we are forced to use cogged wheels. The next requirement is that these wheels shall be as few as possible, and that they should always be in gear. If the variation of ratio is obtained by taking wheels in or out of gear, it can only be done safely while they are standing still and without any load on them. This is equally true with sliding keys or clutches. The next point of importance is that the speed of the driven wheel may be reduced to zero even while the driver is rotating at its full speed. Also that with a driver running at a normal and definite speed, the speed of the driven wheel may be raised from zero to a speed equal to or above the speed of the driver .... The change of ratio to be operated by the simplest possible mechanical arrangement.

Now it is possible to produce these results with two wheels only by using a jointed or flexible shaft; but as this way introduces difficulties, it will be discarded.

With three wheels all the requirements can be easily satisfied, and we may take it that three wheels are the smallest number for producing any ratio. First a driver, second an intermediate, and third a driven wheel.

Rotary motion can be applied to the driver by any usual means, and will not concern us here. The way these three wheels are arranged is shown in Fig. 2. The outside circle is the driver, and the inside the driven.

It will be seen that if the driver were revolving, the driven would revolve in inverse ratio to the number of teeth in the two wheels, and it would give a maximum speed of rotation to the driven wheel. This is so provided the centre of rotation of the intermediate wheel is a fixed point; but suppose the centre of the intermediate is not a fixed point—that it is free to go where it may in the same plane—the result will be very different. The driver will carry it round without giving rotary motion to the driven wheel; in other words the speed of the driven wheel will then be zero.

Here we have the two extremes—the highest speed of driven wheel with a fixed centre of rotation in the intermediate, and a zero speed with a loose centre of rotation in the intermediate wheel. Any control over the centre of rotation of the intermediate wheel will give like control over the speed of the driven wheel, and let it be understood that a control over the centre of rotation of the intermediate wheel is no control over its rotations; it is bound to rotate with the driver, either on its own centre or round the driven wheel.

Mr Chutter controlled the freedom of the centre of the intermediate wheel in a most ingenious way. On the axle of the tramcar he mounted a free running eccentric, shown at F in Fig. 1. This eccentric carried the intermediate wheel running loosely on its circumference. It also was keyed solidly to a brake drum round which a strap could be tightened (shown in Fig. 3). With the brake strap tight, the eccentric remains stationary, and therefore so does the centre of the intermediate wheel and we have top speed. With the brake completely released, the eccentric is free and so is the centre of the intermediate wheel which just tends to rotate round the driven wheel without actually driving it, so the tramcar wheel is not driven either. With the brake strap at any degree of tightness between fully tight and fully free, we have a correspondingly intermediate speed so gradually tightening the strap allows the drive to be taken up smoothly from zero to maximum.

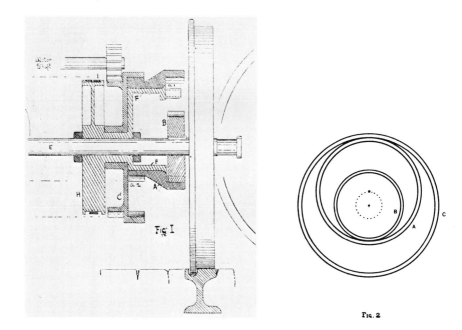

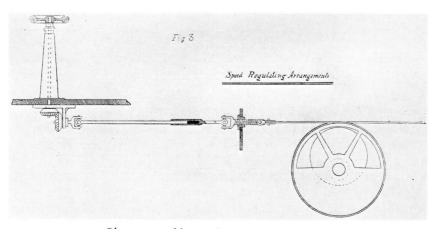

Chutter variable-speed gear for tramcars, 1894

The fly in the ointment so to speak is really a technical one to do with the fabrication of the intermediate wheel, since it has to have internal teeth to engage with the driven wheel attached to the tramcar axle, and external teeth to engage with the driving wheel powered by the engine. Any inaccuracy in marking out and forming the teeth would be disastrous. Wear would also make matters worse. What is more, and is by no means an unimportant point, the intermediate wheel would have been exorbitantly expensive to make.

However, Mr Chutter's gear aroused a great deal of interest and correspondence on the matter—if not to say strong words. A correspondent by the name of Walter Gribben came up with an alternative proposal to achieve the same end but by using only gears with external teeth. These would have been much simpler and cheaper both to make and to maintain in service.

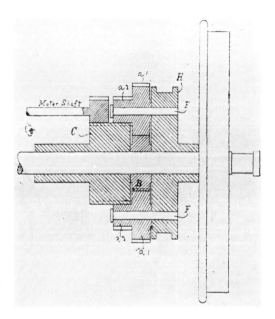

Gribben's variable-speed gear, 1894

In this version, the driven wheel is B and this is keyed to the tramcar's axle. The wide gear C and the brake wheel H are both loose on the axle. H carries with it two steel pins on which rotate a matching pair of cogs engaging with both B and C. It is obvious that tightening a brake strap wrapped round the rim of H achieves exactly the same result as in the Chutter gear. An important difference is that there are no extensive rubbing surfaces to cause wasteful friction as there is between the intermediate wheel and the eccentric in the Chutter gear.

With the increase in the population which took place during the nineteenth century, and the greater concentration of people in towns and cities, it became an urgent problem to provide sufficient satisfactory housing. The long terraces of cottages built to house workers in the various industries are an example of one way of putting up homes cheaply. Some engineers and architects, however, looked at the possibility of prefabricating houses and of using alternative building materials. The firm of Drake Brothers and Reid of New Kent Road, London experimented with concrete cottages in the 1860s. These it was claimed could be put up for £175—ready for habitation. They estimated this to be half the cost of brickwork.

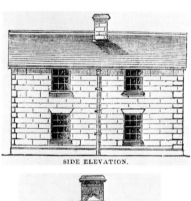

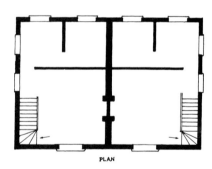

Concrete cottages, 1869

The example illustrated is of a pair of semi-detached agricultural cottages put up on the estate of Sir Arthur Hazlerigg at Illston, Leicestershire. These cottages were basic in the extreme, having only three rooms downstairs and exactly the same upstairs, although really there was only one main room and two little cubby holes. The stairs passed through the main downstairs room. Below the stairs was storage space for coal for the single fire which was the sole source of heat for both warming the house and cooking. The fireplace itself was in the main room, back to back with its adjoining neighbour on the partitioning wall. There were no lavatory facilities in the house and certainly no bathroom. The toilet was presumably a privy at the bottom of the garden.

Naturally too there was no electricity and it is not clear whether piped water was available. If it was, they could have had problems with the pipes freezing up in the winter. Some thirty years later, in 1899, experiments in thawing out frozen water pipes, using alternating electric current, were carried out successfully in Canada. The procedure was to pass a current of about 200 to 300 amps, at somewhere between 20 to 50 volts, through the frozen section of pipe using a transformer connected to the mains supply.

Another novel use for electricity was described in the *English Mechanic* in 1900.

This novel use occurred in an Indiana town, and consisted in the wrecking of a bridge. An old-fashioned wooden bridge had been declared unsafe, and it was determined to remove the woodwork, leaving the masonry intact for the new bridge. The owner of the old bridge agreed to remove it in thirty days, the county authorities having purchased the piers and approaches on that condition. He found, however, that this was no easy accomplishment. He travelled about, consulted bridge and house wreckers, wrote letters, and sent telegrams but all to no purpose. No company or individual was found that would agree to take down the timbers, leaving the masonry intact, in the time available. The 30 days passed and the old bridge still stood. He succeeded in getting an extension of a week, but he was at his wits end. The structure could be blown up with dynamite, but the explosion would destroy the piers also. It could be set on fire, but that would crack or injure the masonry. Several other plans were suggested, but the only sure way seemed to be the erection of false work and that method was out of the question, owing to the shortness of time.

At this juncture a proposition was made to him to remove the bridge by electricity. It was accepted at once and work begun. The method adopted was perfectly simple, and this is the way it was put into execution. Each span of the bridge was composed of nine chords, each consisting of three timbers. Therefore, if these twenty-seven sills were cut simultaneously the span would drop between the piers to the river beneath. That was what was actually done, the cutting being accomplished by burning through the wood by loops of iron resistance wire made red hot by the passage of an electric current and weighted down by sash-weights. The timbers were of yellow poplar and 9 inches square. Each one was burned simultaneously in two places. Thus the mass of timbers dropped well inside the piers without injuring them. It took one hour and forty minutes to wreck each span.

Examination after the fall of the bridge showed that all the sills were burned by the wire loops in exactly the same manner—5 inches deep from the top, and 3 inches deep on the sides. When this depth was reached the weight of the span fractured the remaining wood. The cut made by the hot wire was quite sharp and clean, and the wood was not charred more than an inch from the place of fracture. The current was first turned on about five o'clock in the morning on the day of the wrecking, and at two o'clock in the afternoon the last span crashed to the river bed, and a great shout of admiration went up from the throats of about 2000 spectators who witnessed the feat.

It is just possible that some small child watching the dramatic event could be still alive in 1984 and would be able to recall what it was like. There was no indication in the report of the source of electricity or what the current strength was, so there is no record of how much power was required to wreck the bridge. It was clearly a very neat and relatively safe method.

In 1898 an ingenious plan to use electricity to conquer the world was put forward by the then editor of the *Electrical Age*, Newton Harrieson. The idea was to encircle the earth with a cable and to send sufficient current through it to swamp the earth's magnetic field and produce new magnetic north and south poles along an axis determined by the axis of the cable loop. The original motive behind the plan was to put an end to hostilities between the United States and Spain. By creating an alternative geomagnetic field, Mr Harrieson intended to tilt the earth's axis so that Spain ended up in the arctic region, on the assumption that instant capitulation would be achieved. As Mr Harrieson put it: 'The nation that controls the cable will control the world'.

Quite apart from the physical problems of constructing and laying the cable, or of supplying enough energy to produce the necessary current, Mr Harrieson put aside any consideration of security for the cable. It would have taken more than the entire population of the United States to mount guards on the whole length, day and night. He also seemed to assume that the countries over which the cable would pass would stand aside and let the United States have its way. However, the article itself makes compulsive reading. The following piece is an edited extract.

I would girdle the earth with a giant cable. Through this I would send a current of electricity strong enough to overcome the present magnetic poles of the earth. This would give us a new north pole and a new south pole. The effect of this mighty electric influence would be to transform the earth into a huge electro-magnet of such tremendous power that it would tend to lean violently towards the sun, because the sun's chief constituent is iron.

The only requisite would be the generation of a current of enormous electrical energy. Great turbines would do this. They in turn could operate mammoth dynamos. The fall of water at Niagara would be more than sufficient to operate them. Or the great Zambesi Falls in Africa. In this way power could be generated at a minimum of expense.

It has been estimated that there are 3,000,000 HP stored up in Niagara. This would be much more than my scheme requires. Thus we can see that the United States has all the facilities for carrying out the plan. Think what a power this country would become in adjusting the affairs of the world. Expense? Nonsense! It would be a trifle in comparison with the benefits that would result. Look at the cables that have already been laid throughout the world. The work in doing this has been far greater than the work required for laying my giant cable. There would be no trouble in laying this girdle over land and sea.

In the universe the earth is but as a tiny ball of pith, light as a feather. It is most delicately poised in space. The slightest force is all that is needed to disturb it as it dances before that mighty ball of electricity—the sun. The sun is 1,281,900 times greater than the earth. It is as an orange to a pinhead. An insignificant force will change the position of the pinhead.

The cable laid, the current turned on—let us consider the benefits of the change in the position of the earth's axis. We could save all the vast crops that are annually destroyed by the heat or cold. We could reclaim all the vast stretches of land that heat and cold alike render barren. The cold in the Arctic regions is not due to distance from the sun. As a matter of fact we are nearer the sun here in the Northern Hemisphere in winter than we are in summer. The angle at which the sun's rays strike the earth is the sole factor in determining heat and cold. I change that angle.

Bring the cold parts of the earth more directly under the balmy rays of the sun, and, presto! we have summer where winter reigned. Greenland would blossom with fruits and flowers, wheat fields and gardens would take the place of ice fields and glaciers. So in the Tropics the fearful heat might be tempered. Gone would be yellow-fever in Cuba, orchards would take the place of the jungle. Any spot on the globe could be reclaimed and rendered habitable. The value of the land thus converted from barren fields of ice into productive fields can be scarcely estimated. The continents within the Arctic Circle comprise thousands of square miles of land provided with natural waterways and in every way fitted to support a large population.

Think what could be done in war times! The nation in control of the current could annihilate an enemy at will. Such an attack would be beyond the power of any fortifications or any guns to withstand. In a single night a nation could be wiped off the earth—frozen to death. If the United States controlled the cable it would be the dictator of the world.

Warfare would have to disappear in the face of such a terrible agent of destruction. The nations would have to agree or die. If they did not they would have to face certain destruction. The United States at a single touch of the button would transform them in turn to frozen wastes or torrid deserts. The waterways could be made to freeze or boil. In an hour the ocean would be a solid mass of ice or seething with heat. No life could withstand the change. Ours would be the victory, without the loss of a single life.

Aside from these terrors, see what a benefit such a plan would be. Man could absolutely control the seasons. He could cool the Tropics and warm the Arctic and Antarctic. The whole world could be made like a beautiful garden with proper attention. All the earth's surface could be reclaimed for agriculture. Bread famines would cease. Crops could be raised everywhere without fear of loss by varying temperature.

And everybody would be happy.

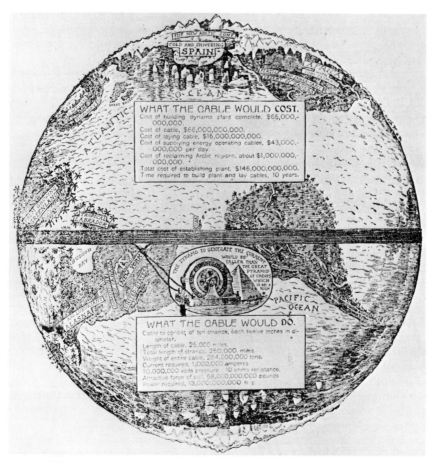

Harrieson's method of creating an alternative geomagnetic field, 1898

A month or so after Mr Harrieson's article appeared a rather more detailed analysis of the power needed to move the earth was reported in *La Nature* as follows:

Statisticians sometimes have queer ideas. One of them has amused himself by calculating how much energy, water and coal it would take to move the earth a foot, supposing that it was subjected throughout its mass to a force equivalent to terrestrial gravitation. This is a gratuitous supposition, for in spite of its enormous mass the earth *weighs* nothing, and it is only by piling up hypotheses that we can get an idea of Archimedes's famous lever. Starting with the fact that the earth's mass is about 6,100 million-million-million tons, our statistician calculates that we should require 70,000 million years for a 10,000 horse-power engine to move our globe a foot. The boiler that should feed this engine would vaporise a quantity of water that would cover the whole face of the globe with a layer 300 feet deep. The vaporisation of this water would require 4,000 million-million tons of coal. This coal carried in cars holding 10 tons each, and having a total length of 30 feet, would require 400 million-million cars which would reach 80,000,000 times round the earth. This train moving at the rate of 40 miles an hour would take more than 5,000,000 years to traverse its own length. It would require for storage a shed that would cover a thousand times the area of Europe. If we realise that this fantastically huge amount of energy is as nothing at all compared with what the earth possesses in virtue of its rotation about its axis, its revolution about the sun, and its translation in space with the Solar System, of which the earth is but an infinitesimal part, and which itself is but an infinitesimal part of the universe, we may get some idea of the importance of man in the universe, and estimate his incommensurable pride at its just value.

Still on a military theme, the use of electromagnetism in a more 'conventional' way was proposed by Nikola Tesla in 1898. The date is significant because what Tesla designed was a radio-controlled torpedo, or 'dirigible boat' as it was called. The intention was to be able to steer the torpedo to its target either from a safe point on shore or from a distant ship, by means of 'electrical waves'.

Again this was put forward as a weapon that would 'abolish war'. He claimed that 'War will cease to be possible when all the world knows tomorrow that the most feeble of the nations can supply itself immediately with a weapon which will render its coast secure and its ports impregnable to the assaults of the united aramadas of the world. Battleships will cease to be built, and the mightiest armourclads and the most tremendous artillery afloat will be of no more use than so much scrap iron'.

For some reason Tesla could not resist a dig at the British Navy in his enthusiastic description of what his torpedo could do. He claimed that 'England is now no stronger than the weakest of the maritime nations . . . She will be utterly confounded, . . . and France will rejoice'. He even went so far as to state that 'it could strike a vessel that lay at Southampton, England, while the operator was snugly ensconced in the forts at Sandy Hook'.

The important thing about the date of Tesla's invention is that it was only the previous year, 1897, that Guglielmo Marconi had set up the first radio transmitting and receiving station in Britain, at Lavernock Point, Penarth, near Cardiff. It had taken a mere decade from Heinrich Hertz's experiments with an induction coil spark transmitter and a wire loop receiver in 1887 to Marconi's successful transmission of a message in Morse code from Lavernock Point to the little island of Flat Holm 3 miles away in the Bristol Channel. In December 1901 he succeeded in transmitting a message across the Atlantic.

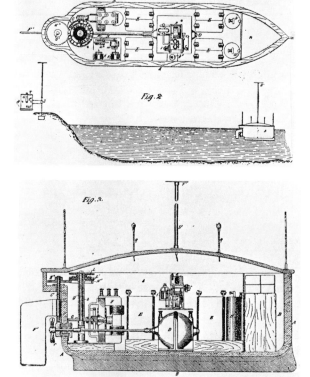

Tesla's dirigible torpedo, 1898

In the United States, Thomas Alva Edison too had turned his attention to wireless telegraphy and obtained a patent for his system in 1892. In his specification he says:

I have discovered that if sufficient elevation be obtained to overcome the curvature of the earth's surface and to reduce to a minimum the earth's absorption, electric telegraphing or signalling between distant points can be carried on by induction without the use of wires connecting such distant points. This discovery is especially applicable to telegraphing across bodies of water, thus avoiding the use of submarine cables, or for communicating between vessels at sea and points on land.

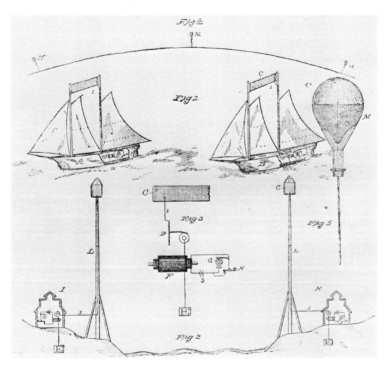

Edison's method of telegraphing without wires, 1892

Edison's transmitter was a rather curious device which would not have lent itself to clear communication using the Morse code. It consisted of the usual induction coil with one terminal of the secondary coil connected to an aerial mounted either on a tall mast or supported by a captive balloon, and the other terminal earthed. The primary circuit contained, as well as a battery, a rotating circuit breaker driven by a small motor. The circuit breaker was kept short-circuited by a key. Only when the key was pressed was the circuit actually 'activated' but depending on the speed of rotation of the circuit breaker, at each press of the switch there would have been several impulses in succession. Rapid signalling would have meant a difficult job at the receiving end to tell when one burst of impulses ended and the next began.

A military gadget of no practical value at all was invented in 1896, in good time for the Boer War, by a Mr Aitchison. It consisted of a normal pair of high-powered binoculars mounted on a special frame which could be strapped to the head. The intention was to enable an observer to see the details of enemy positions, fortifications and so on, but leaving the hands free to make sketches. The trouble is that although distant objects could be seen very well, the observer couldn't see what he was actually drawing on his sketch pad. He was no better off than if he had held his binoculars by hand. . . .

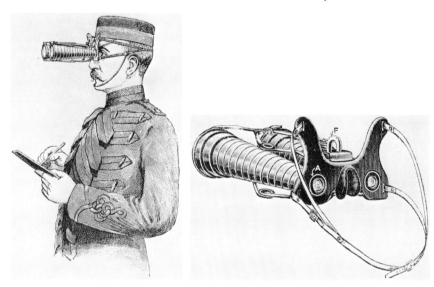

Aitchison's patent head-rest for field glasses, 1896

Turning now to something completely different, it is worth reflecting on how useful it is to be able to make multiple copies of documents cheaply. We now take the familiar photocopying machine for granted, not just as a convenience but as an essential piece of office equipment. In 1877, Edison was granted a patent for what he called an 'electric pen'. This was really the ancestor of the modern duplicator. His initial attempts to devise a mechanical copier led to the perforated stencil made by writing on a specially prepared kind of paper with a stylus, the paper being placed on a steel plate whose surface contained thousands of fine grooves. This process caused the paper to puncture with hundreds of tiny holes in the outline of the writing. When the paper was inked with a roller, the ink was squeezed through the perforations onto blank sheets, making an exact copy.

The next step was to have a stylus with a vibrating point, moved up and down by means of a tiny electric motor. Stencils made by the Edison electric pen could produce up to 3000 legible copies. A new kind of electric pen was invented by Messrs Ballet and Hallez d'Arros in 1879 in which an electric spark caused the perforations in the stencil rather than a vibrating steel needle point. The device was described in *La Nature* as follows:

The spark of an electric machine or an induction coil passing between metallic points or between a point and a conducting body is capable of piercing a card, and will, of course, much more easily puncture a sheet of paper.

When the sheet of paper rests on a metallic plate and the surface is traversed by the electric pen, the plate and the pen being connected with the poles of an induction coil, a line may be produced by a series of very fine perforations, which will vary in number in a given space with the rapidity of the discharges and the rate of movement of the pen. The principle of the pen is very simple, but before the practical utilization of it was reached, many difficulties had to be surmounted. Among these may be mentioned the tendency of the sparks to burst forth when the pen is within a short distance of the paper, puncturing the paper in all directions, making it impossible to draw a clear line from the start. The operator was also liable to severe shocks.

Another difficulty was the distance between the successive perforations. These imperfections have been overcome ... by reducing the strength of the secondary current, so that it has only sufficient power to pierce the paper, and will not, therefore, give a perceptible shock.

The paper which is to form the stencil is dipped in a solution of salt and dried; this operation prevents too many sparks from issuing from the pen, and insures an absolutely true and clear line. The interrupter is of novel form and is operated by the magnetised core of the induction coil. The apparatus forms a desk of medium dimensions. At one side of the desk is a plunging bichromate battery; the induction coil is placed in the middle and is connected by one of its wires with the lead of an ordinary lead pencil, which serves the double purpose of making a visible mark on the paper and of conducting the current. The metallic plate which supports the paper is also connected with the coil and is secured to the desk top. When it is desired to take an impression from the stencil, it is placed over a sheet of paper and rolled with printer's ink reduced with a little printer's varnish or with castor-oil.

While on the topic of office work, another idea which is usually thought of as a modern invention is 'flexitime'. This is a system which gives flexibility in the hours that staff actually spend in the office. The agreed number of hours per week can be distributed, within reason, to suit individual convenience. The original idea started in the United States at the International Correspondence School, Scranton, Pennsylvania as a means of preventing congestion on stairways and elevators in the building by staggering the times of arrival and departure of people working in different offices. This report appeared in 1900:

The time of entering and leaving the building is regulated by clocks on each of the five floors. On the lower floors the clocks are set correctly, but on the upper floors they are a few minutes slow, so that the employees on the lower floors are at their desks before those on the upper floors are due at the building. In leaving the building the employees on the upper floors do not leave their desks until several minutes later than those on the lower floors.

To have the clocks set at different times seems unnecessarily complicated and unrealistic unless the employees set their own clocks and watches to match, After all it would have been simpler to have the staff on the top floor work, say from 9.15 to 5.15, while those on the ground floor worked the conventional 9 to 5.

We cannot leave the world of work without saying something about holidays—after all a holiday is not a holiday unless you are usually at work. Anyone planning a holiday in 1898 might well have looked up 'Hints for holiday makers' in the *English Mechanic* where suggestions were made as to what was essential medication to be carried with all the other luggage.

The medicine chest when prepared for the country should contain large bottles of ammonia, carbolic acid, vaseline, glycerine, witchhazel, arnica, and glisterine, some simple disinfectant—permanganate of potash is good—a package of powdered borax, and one of pure cinnamon. Dr Duncan says: The last will be found invaluable if you chance to be in a neighbourhood which is badly drained and there is a danger of typhoid fever. It should be steeped and taken freely as a drink; for it has the power to destroy infectious microbes. Even its scent kills them, while it is perfectly harmless to human beings.
    For incipient scratches and slight cuts, no more healing lotion can be found than glycerine containing a few drops of carbolic acid. It allays all pain and smarting instantly. If the bones are weary and strained from tramping and climbing, all the joints, the backs of the calves, and the thighs should be well rubbed with vaseline taking if possible a warm bath first. After a good night's rest you will feel fresh and ready to start for another day in the open.
    Be sure to sleep in a well ventilated room and don't be half so afraid of draughts as of being deprived of your necessary allowance of fresh pure air. Don't burn a night lamp unless sickness renders it indispensable; under no circumstances allow a turned down paraffin lamp to pollute the atmosphere.
    In choosing your abiding place for the summer make plenty of sunlight a condition. Insist upon sleeping-rooms that are daily purified by the sun's

rays. North rooms may sound cool and attractive, but remember always that they cannot fulfil the conditions of perfect hygienic living.

After a fatiguing tramp, the tired body should be prepared for restful sleep by a careful night toilet. If there is no convenience for a plunge bath, the body should be sponged off with warm water containing a few drops of ammonia—if the feet can be left in the foot bath for ten minutes it will be all the better; rub very thoroughly with a Turkish towel, and, last of all, refresh the face, neck and arms by spraying with rose water, toilet-vinegar, or any favourite toilet-water. Brush the dust out of the hair, and wipe it with a towel; gargle the throat with salt and water and clean the teeth. Sweet restful sleep should follow this regimen, and prepare you to waken on the new day, fit, mentally and physically for any duty or pleasure that awaits you.

One can't help but notice that even on holiday 'duty' comes before 'pleasure' in the Victorian scheme of things. However, before rushing off to the railway station for a ticket, it might be prudent to consider a warning about the dangers of holidays, and the waste of valuable time that they entail, that appeared in 1896 in the *Medical Press*.

It is one of the fashions of the times to assert that holidays are more necessary now than they were not so many years ago. The reason usually alleged is that, owing to the high pressure and hurry of the present day the human brain requires longer and more frequent rests than formerly, and that competition is so great that a larger number of 'days off' are absolutely necessary to repair the waste of grey matter used up in the inevitable struggle. We are inclined to think, however, that the holiday craze is going too far. The best mode of giving the brain tissue its required rest is not to indulge in furious 'biking' nor yet to drowse away a week or a month in a sleepy hollow. The brain does not need, when healthy, even a week's rest: a good night's sleep is much more to the purpose. Still better is a hobby, and especially one which calls for some mental effort different from that required in the daily work. Any professional man who has no interests beyond his profession, or no chance of varying his daily duties has our sincerest pity. We are not at all sure that holidays of more than ten days or a fortnight at a time are good for anyone who is in good health and has work to do. For those who never do any real work all time is practically a holiday and it is immaterial when it is spent.

The return after a long holiday is usually signalised by restlessness, inability to concentrate the mind to the details of work, and, though this may appear paradoxical, by a proneness to attacks of disease. It would be quite worthwhile for someone to investigate the statistics bearing on this point. Short holidays two or three times a year are probably of more use than one long one, while if the weekends are often out of town, less than that is enough. The constant wish to get away from work, which is so characteristic of the present day, indicates little love for it, and that betokens degeneracy ...

A commonplace experience of present day holidaymakers, especially when taking holidays abroad, is to have one's luggage inspected. It is perhaps surprising to find that in 1897, only two years after Wilhelm Conrad Röntgen made his discovery of x-rays, a M Seguy at the School of Pharmacy in Paris developed a portable x-ray machine that enabled customs and excise officers to detect arms and other metal objects in travellers' baggage.

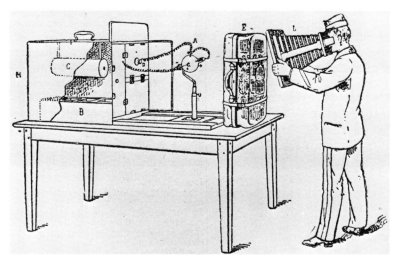

Seguy's portable x-ray machine, 1897

The illustration shows how the machine—called a 'lorgnette humaine'—was set up and operated. The officer operating the machine was given no protection from the x-rays, yet he stood only about half a metre from the x-ray tube.

An example of how developments in science and technology can sometimes bring unexpected—and unwanted—results was reported in *The Lancet* in October 1898. The article describes an accident in which a woman pedestrian was killed when she was run over by a horse-drawn coal wagon in London. The cause of the accident was an illuminated advertising sign. The sudden flashing on of the light startled the horse and dazzled the unfortunate woman, with tragic results. The jury at the ensuing inquest lodged a protest against the use of flashing signs, a sentiment with which the writer of the article agreed.

... it seems worthwhile to point out with greater fulness the nature and extent of the dangers which may arise from these inventions. These advertisements are often lighted by powerful electric lights, and during the 10 or 20 seconds in which the lights are turned on they brilliantly illuminate the part of the street in which they are placed. Suddenly this brilliant light is turned off, and the streets are left in that condition of semi-darkness which the nearest gas-lamp is wont to produce. The result of this sudden transition is obvious. If we suddenly pass from brilliant sunshine into a dimly-lighted room it is a matter of common knowledge that for the first half minute or more we are quite unable to see anything. The eye takes an appreciable time to accommodate itself to the altered conditions. So it is with these most dangerous advertisements. For a few seconds everything is as bright as day. Then in a moment we plunge into what appears to be complete darkness. If we are so unfortunate as to find ourselves in the middle of the roadway when the light is turned off we have an excellent chance of never reaching the pavement except on a stretcher. The foot-passengers cannot see; the drivers cannot see; the horses cannot see, and, moreover, they are often rendered nervous by the sudden change from light to darkness. ...
We should like to have the opinion of the policeman on duty at some point where a flash-light advertisement is in operation as to the additional difficulties which it causes him. A constable on duty, say, at the top of Sloane Street has none too easy a task in regulating the traffic even when all the conditions are favourable and he has a steady light to see by. But place him at intervals of 20 seconds in the full glare of a powerful electric light and then in what appears by contrast to be total darkness and he is practically helpless. A man cannot regulate the traffic when he cannot see it. And yet this very point, the top of Sloane Street, a narrow and very crowded thoroughfare where several lines of traffic converge, has been selected for one of these electrically-lighted advertisements. It seems to us astounding that such a state of things should be permitted. It should be possible to find methods of advertising a tobacco, a beef tea, or a patent medicine which are not actually dangerous to the safety of the community ...

Some paper cannon have recently been completed for the German army, which, however, are not expected to replace those of metal. They are merely intended for use in portions of the field where the taking of metal guns is impracticable.

English Mechanic 1898

AN ENGINEER'S EPITAPH

My engine now is cold and still,
No water doth my boiler fill,
My coke affords its flame no more,
My days of usefulness are o'er,
My wheels deny their wonted speed,
No more my guiding hand they heed.
My whistle too has lost its tone,
Its shrill and thrilling sounds are gone,
My valves are now thrown open wide,
My flanges all refuse to guide.
My clacks also, though once so strong,
Refuse their aid in th' busy throng.
No more I feel each urging breath,
My steam is now condensed in death.
Life's railway's o'er, each station's past;
In death I'm stopped, and rest at last.

This epitaph is on the tombstone of Oswald Gardener who was killed when the locomotive *Wellington* he was driving broke a connecting-rod at Stocksfield in 1840. He was buried in Whickham church, near Newcastle-upon-Tyne.

Chapter 2
Domestic Science and Practical Household Economies

IT is difficult now for us to really appreciate what a hard life ordinary people had years ago. Much of the gadgetry and other conveniences that we take for granted, almost without question, such as vacuum cleaners, refrigerators and freezers, washing machines and dish washers, convenience foods, electric or gas cookers and many others have only come within the reach of the vast majority since the Second World War. A hundred or so years ago these things either didn't exist, or if they did they were only within the reach of the very rich. Moreover, while domestic labour remained cheap and plentiful, there was little incentive for change and those who would have appreciated a lessening of the burdens of household chores couldn't afford labour-saving gadgets anyway. In the United States, however, circumstances were rather different to those in Britain. There labour was more expensive, and in a much less class-ridden society there was a greater incentive to find easier ways of doing things.

One of the most unpleasant chores that had to be faced was the weekly wash—except for those who could afford to give the work to a private laundry company. Mitchell's steam washer of 1883 offered a very rapid service, being able to do a fortnight's washing within an hour—an achievement all the more remarkable when one considers the speed at which modern automatic washing machines work. It was described as follows:

Mitchell's steam washer—like most washing machines, consists of a revolving cylinder inside a rectangular box with oval lid. The whole machine is made of copper, and the clothes are put inside the cylinder. Enough water to make steam is put in the box, and by means of a Bunsen burner, a common fire, or steam direct from a boiler, the water is made to boil rapidly. A lever keeps the clothes continually in motion, and in ten minutes or thereby, all are clean. It is claimed for this machine that the water on the clothes in being turned into steam blows off the dirt. This may be true but engineers are of the opinion that the peculiar fluted construction of the inside of the cylinder also aids in producing the marvelous work of accomplishing a fortnight's washing in an hour.

This machine sounds amazingly modern—a rotating drum inside a rectangular box. However, what the report doesn't point out is that the mechanical movements had to be accomplished manually. Strangely too there is no mention of soap being used. What is more it must have been a tricky job to load and unload the machine with clothes while it was still hot—especially if it was heated on 'a common fire'.

A machine designed to rapidly force the washing liquid many times through the clothes 'with the least possible expenditure of labour or power, and without danger or injury to the clothes' was patented by William Acheson of No 2307, Penn Avenue, Pittsburg, Pennsylvania in 1896.

Acheson's washing machine, 1896

The cylindrical clothes-receptacle has in its periphery a removable cover, through which are introduced the washing liquid and the clothes to be washed, and its heads have hubs which turn in bearings on suitable standards. The water is forced through the clothes by reciprocating perforated plungers or dashers whose squared shafts slide in and turn with the hubs, there being on one of the hubs a pulley to be connected by belt with a source of power, or the machine may be operated by hand through a gear-wheel on the hub which meshes with another gear-wheel actuated by a crank. The reciprocating motion is given to the plungers by double cams on the outer ends of the plunger shafts, the cams engaging friction rollers to give inward impulses, while the return motion is effected by springs coiled on the shafts.

By the look of it this machine must have weighed a ton. The interesting thing about it though is that it could be power driven. On a farm where power was available this must have been a considerable boon. Elsewhere, however, the massive construction and the peculiarity of the action would have made manual operation very exhausting. How the water was heated too is not explained.

A more sensible-looking machine of modest proportions was reported in the *Scientific American* in 1897. This was the invention of Fred R C Pitzler of Lester Prairie, Minnesota. This machine could be turned on its back for loading and unloading, but was tipped on its side when it was being activated.

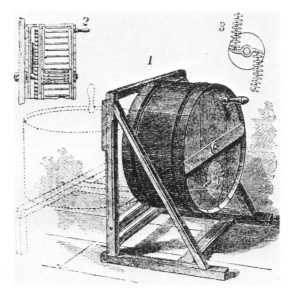

Pitzler washing machine, 1897

Another American machine, this time of 1900 vintage was patented by Fridolph and Minnick of Villisca, Iowa. This one was designed to fit on top of an ordinary cooking stove.

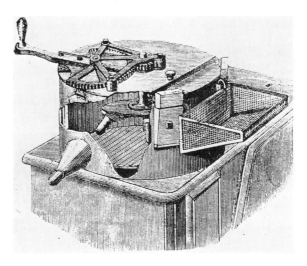

Fridolph–Minnick washing machine, 1900

The particular advantage of this machine—apart from utilising heat that was already there in the stove—was that the clothes were held in an inner basket and agitated by a backwards and forwards rotating paddle, the dirt falling through the spaces in the sides and bottom of the basket and being thus separated from the clothes. The 'sludge' could be run off through the spout shown on the side of the drum, normally closed by a plug. The washed clothes could be lifted out into a wire tray and a wringer could be attached to squeeze them dry.

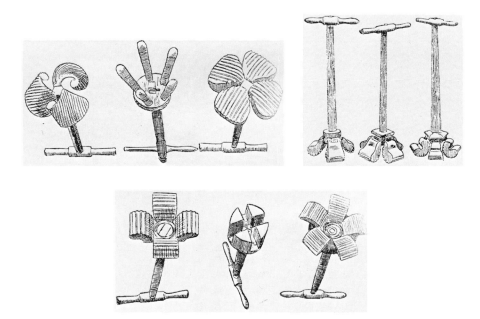

A selection of peggies, 1897

Those who could not afford a washing machine either had to do the laundry literally by hand, or by means of a specially shaped wooden rubber called a dolly or a peggy. Usually these were used in conjunction with a tub having ribbed sides for holding the water and the clothes. As a small boy I frequently watched the skilful way that my grandmother operated her peggy in the dolly tub. The idea was both to thump the clothes and to rub them between the ribs on the inside of the tub and the specially shaped prongs or rubbers of the peggy. These were still available in the 1940s. Cheap versions had peggies made with rigid wooden rubbers, while more sophisticated ones had the rubbers attached to the handle by crude hinges making them less damaging to the clothes by being more flexible.

An article in the *Morning* in 1897 by a Mrs Humphry discussed at length the problems associated with obtaining well trained and reliable servants, and showed how going for the cheapest meant false economy.

'Thrift,' said Sir Philip Sidney, 'is the fuel of magnificence' and it is ever so many things besides. We do not all care about magnificence. Our ambitions run more towards nearer goals, ease from money cares, comfort for our old age, opportunities for travel, facilities for reading the best of current literature, for hearing good music and enjoying good pictures, to say nothing of the delightful capacity to help others a little bit when they droop and fall by the wayside, fainting on life's toilsome journey. Thrift helps us to all of this, and delivers us from the perpetual struggle of carking care, which darkens life for those who live up to and beyond their income.

But by thrift we do not understand anything approaching parsimony, do we? Not in the least. Enough for everybody, but nothing wasted is what we mean. Not even time! Early breakfast, for instance, economises the time of the whole household to an astonishing extent. No one could conceive how much without practical experience. And there are methods of economy that initially cost more, but eventually save surprisingly. Take the cheap servant girl for instance. We can get her for £10 a year, perhaps, whereas an experienced maid would cost at least £17. So! We save £7 a year. But, if you please, how much do we lose? Let anyone who has leisure try to estimate what that 'cheap' girl costs us in pounds, shillings and pence, irrespective of worry of mind and irritability of mood.

The Cheap Girl dabs down a very dirty duster on a clean quilt, wipes the looking glass with a greasy cloth, hangs up one's best gown somewhere by the back of the neck, thrusts a black hand into the boots she cleans, leaves pails on the stairs for people to fall over, smears the cups and saucers instead of washing them, leaves empty stew pans and kettles on the range, and conseqently burns the bottoms out of them, puts saucepans away dirty, gets the sink stopped up, sets the dustbin on fire by putting hot cinders into it among the waste paper, leaves strangers in the hall while she searches for her mistress, and thus gives them opportunity for theft, forgets to give letters and messages, and in a score of ways worries and irritates her unfortunate employer. Her breakages too are a serious matter to the mistress of a small establishment and a narrow income. All these things make the cheap girl a remarkably expensive addition to the household. Far better to give higher wages and secure one who has been in some degree trained, who knows how to handle glass and china, and will not treat them as if they were lead or brick, and understands how to take care of pretty furniture, and to be neat in her dealings. Far better have one good experienced servant, even if her wages come to as much as those of two cheap girls. Suppose that a young wife is fortunate enough to get a thoroughly competent young woman to do the whole work of the house. To pay her well is an incentive to remain, and if she is economical and trustworthy, she is well worth £25 a year. She cannot eat as much as two would, and if she is well paid she soon settles down, and makes her mistress' interests her own.

When one comes to reckon up the cost of burst boilers, cracked kettles and stewpans, broken china and crockery, food ruined in the cooking, repairs to cisterns, taps and furniture, necessitated by clumsy handling, the £7 a year we are supposed to save in wages are soon swallowed up. And there is another item to be included in the charges against the ordinary low-waged girl. She eats enormously; seems never satisfied; and the cost of her food is much greater than that of the full-grown domestic whose growing days are over. Consequently, she is by no means an economy, notwithstanding her low wages. Her consumption of food combined with her lack of experience render her an extremely costly addition to the household, and after three months of her there is hardly a whole teacup or saucer and scarcely an undamaged teapot left in the place.

I have always found that the poorer a woman is, the more wasteful are her ways. The best servants come of thrifty, well-to-do families of the tenant farmer, or small shopkeeping class. They have been taught economy in little things all their lives and the lesson has sunk deeply in. The daughters of the very poor, on the contrary, have lived from hand to mouth, with never a sixpence in tenure beyond a day, and no settled income or expenditure—such as these are the most wasteful of human beings.

Apart from wasteful servants, another unwanted menace in the household was vermin. The *Bulletin du Photo Club de Paris* carried a report in 1896 showing how rats had been eliminated from the Chateau Dobroslawitz in a most economical manner:

Count W—— is an enthusiastic amateur photographer, and has trained some of his servants to undertake some of the minor functions of the laboratory assistant. One of the valets was in the habit of leaving candle ends about and remnants of food. This attracted the rats from the stables and caused much trouble and annoyance. Infallible traps were ordered from Paris, but to no purpose. Phosphorus paste and other poisons were not to be had; but a visitor, finding a stuffed cat, a former favourite of the household, he thus ingeniously defeated the wary rodents. The eyes of the cat were removed and replaced by a couple of lenses from a toy microscope. The Count's secretary placed two small incandescent lamps behind the lenses. These were connected with a battery and the cat, thus equipped, was placed in the laboratory near the suspected entrance of the rats. A very delicate connection was made with the lamps and covered with a cake made from flour, melted fat, oil, and powdered sugar. This was placed in such a manner that a rat seizing it would immediately put the current into pussy's eyes. The Count had no faith in the arrangement; but the watchman undertook to keep an eye on the laboratory, and the first night the cat won a complete victory. The rats had scarcely touched the cake when the eyes of the cat were illuminated with a greenish light, and terrified the devastators so utterly that they fled stupefied and never returned again. An electric bell was rung at the same time as the eyes were illuminated. The watchman was immediately on the spot and found the cake partly gone, and the eyes of the cat still glowing with their deathly light.

Insects of various kinds were another common form of household pest. The accepted treatment for cockroaches, for example, was to put down at night in places where they tend to run, little heaps of a substance known as Persian powder, or pyrethrum roseum. The experience of a man in Shropshire, in 1900 makes interesting reading.

Some 25 years ago I took a house where blackbeetles abounded. I bought a spraying apparatus for one shilling and another shillingsworth of powder for refills. The first morning's catch was about a pint of beetles, dead or stupefied, which were scalded (not fried in butter for breakfast as they are in some countries). The second morning about the same number were bagged, but after that the decrease was rapid and in about ten days not one was left, nor did any reappear.

While on the subject of pests, we could perhaps include burglars. Here is a very elaborate burglar alarm devised in 1869 by an engineer named Bertie Parfit.

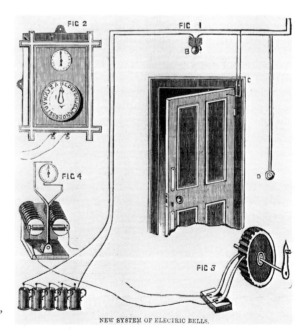

Parfit's burglar alarm, 1869

I have been unfortunate enough to have my house entered by burglars, consequently I determined to fit up the alarm bells which would connect every door and window after nightfall, as I have an idea that should any other unwelcome intruder be stealthily prowling round the premises, it would be much better for him to spring the alarum and make his exit, than for me to have an encounter with him.

As some of our numerous subscribers think so too, and may wish to do likewise, I will describe how I have proceeded. I have made a sulphate of mercury battery because it has been described as the constant battery, and is said to remain in action for twelve months; however, this has to be proved. It has, however, been in action two or three months and continues to work very satisfactorily. I think much may be said in favour of this description of battery; it is very simple and effective, no porous pots required, no unpleasant fumes from acids &c. I shut it up in the cupboard out of the way and out of sight; it requires no attention. Two wires, one from each pole, run through every room, strained tightly on the walls a few inches from the ceiling, and made secure by a tin-tack, taking care the wires do not touch or come into contact with anything that would conduct the electricity to earth, then the doors and windows are connected to these wires in a very simple manner, and the whole thing is almost invisible when the door is shut. A few inches of silk thread above the top hinge inside pulls the ends of two wires asunder about an eighth of an inch, but directly the door is opened the thread is slackened and allows the wires to make the connection and all the bells in the circuit will ring as long as the door remains open . . . to prevent unnecessary bell ringing during the day I have a switch in my bedroom to put on at night.

I have also other switches which control the electric current by day so as to make the bells available for domestic use by the ordinary press buttons.

Having got thus far Mr Parfit came up against a problem he was unable to solve. His daughter became ill and confined to bed for a time and her bedroom was up two flights of stairs. He had the idea of using his burglar circuit for operating an alphabetic telegraph system and he succeeded in making a suitable transmitter (shown in Figs. 2 and 3) so that she could signal to anyone in view of the receiver (Fig. 4). Unfortunately he couldn't make it work. The transmitter could be hung on a wall at some convenient place and operated by moving a pointer to each of the letters of the alphabet in turn arranged round the dial. The signal from each letter was transmitted through a rotary switch (Fig. 3) made of wood, with 26 conducting segments corresponding to the letters. He appealed for help from knowledgeable readers of the *English Mechanic* but, sad-to-say, no-one was able to come up with any practical suggestions. However, he only had to wait until after 1876 when he could have installed a telephone which would have solved the problem admirably.

An alarm of a slightly different kind was reported in the *Scientific American* in 1899 as the invention of Arthur De F Risley of Richfield Spa, New York.

In this case the normal carpet underlay was replaced by one which had conducting strips on its upper and lower faces. Pockets in the underlay contained lead pellets in contact with the lower conductors.

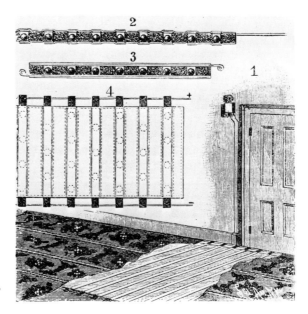

Risley burglar alarm, 1899

Anyone treading on the carpet would cause the underlay to compress and bring the upper and lower conductors into electrical contact through the lead pellets, thus closing the circuit and ringing an electric bell hidden behind the door. When the pressure was removed, the natural springiness of the underlay separated the contacts again and switched off the bell.

In March 1876, Alexander Graham Bell was granted a patent for his now universally known device that made it possible for the first time in history to send messages through wires over long distances using the human voice. The conventional telegraph became obsolete almost overnight. However, Bell's original telephone was far from perfect—especially in the transmitter part. The first major improvement came in 1877 when Thomas Alva Edison received a patent for his carbon button microphone transmitter.

In the Bell system, words spoken into the mouthpiece caused a soft iron diaphragm to vibrate in close proximity to one of the poles of a bar magnet. The end of the magnet nearest the diaphragm was surrounded by a coil of fine insulated wire and the fluctuations in the magnetic field caused by the vibrations of the diaphragm induced a weak fluctuating current in the coil and its circuit which of course included the line to the distant receiver. These fluctuations corresponded to the sound waves in the spoken words so that in effect it was the voice which generated the current; a loud voice meant a good strong signal! In practice this really meant that the system was only practical for short distances. The receiver was an exact counterpart of the transmitter and the incoming signal generated a fluctuating magnetic field which vibrated the diaphragm to reproduce the originating sound pattern. In fact, the receiver served as the transmitter too. The telephone user spoke into the mouthpiece then placed it to his or her ear to hear the reply.

The Edison system, on the other hand, required a 'carrier' current to be supplied by a battery in a primary circuit linked through an induction coil to the secondary circuit which contained the line to the distant receiver. The microphone transmitter contained a carbon button in light contact with a diaphragm. Vibrations of the diaphragm interrupted the primary current which then induced a fluctuating high voltage current in the secondary circuit. The much higher voltage meant both a stronger signal and a far longer transmission distance than in the Bell system. At the receiving end another induction coil reduced the voltage to that required to operate the earphone which was identical to that of Bell. Another advantage of the Edison system was that the receiver and the transmitter were separate instruments and this meant a more convenient arrangement and method of operation.

However, as with most things, there were always people who knew better and the early history of the telephone is one of bitter argument about prior claims, litigation and dispute. The century since 1876 has seen tremendous developments in telephone technology where it might have been thought it had reached perfection long ago. Sometimes these developments seem to have been backwards instead of forwards, as for example the telephone system devised by a Mr Andrew Plecher of Stanford University, California. This appeared to combine in one instrument the worst features of all the alternatives that had been tried. This report, which was extracted from the *Scientific American*, appeared in the *English Mechanic* just before Queen Victoria died in 1901:

A new telephone transmitter and receiver has been devised by Mr Andrew Plecher of Stanford University, the peculiar construction of which is shown in the accompanying diagrams. Of these diagrams Fig. 1 is a view of two combined transmitters and receivers; Fig. 2 is a slightly modified form.

The transmitter and receiver consists of an iron box M, connected by a heavy iron wire I with a similar iron box. In each iron box there are two thin diaphragms, D and D', insulated from each other by a non-conducting marginal ring N forming an airtight joint with the diaphragms. Behind the diaphragms, in the box M is an opening O for the admission and discharge of sound waves. In the hermetically sealed chamber, thus constituted a coil X of fine wire is suspended, so wound that the individual turns nearly touch one another. One end of the coil is connected with one diaphragm D and the other end with the diaphragm D'. When an electric current passes through the coil the turns will touch, since the coil becomes magnetic. The vibrations of the diaphragms will separate or bring into contact the turns of the coil, whereby resistance is thrown into or out of the circuit, thereby causing

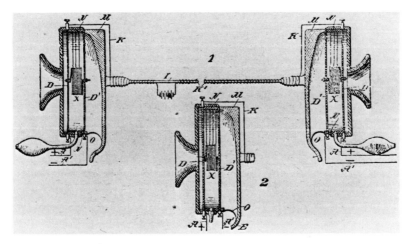

Plecher's electro-pneumatic telephone, 1900/1901

a corresponding fluctuation of the current. For the magnetic action of the current causes the turns of the coil to be attracted. Then when the vibrating diaphragms move outwardly his lateral contact is broken and the resistance of the whole coil will be thrown in by compelling the current to traverse the coil lengthwise instead of leaping from turn to turn. The air vibrations propagated by the voice act on the front face of the diaphragm D through the mouthpiece and on the rear face of the diaphragm D' through the opening O, whereby the two diaphragms are caused to vibrate in opposite directions. The effect on the resistance-varying coil X is thereby augmented. The fluctuations are transmitted through a circuit composed of a fine wire K connected with the diaphragm D and wound round the central stem of the box and the heavy wire I, converting the box and the wire I into a magnet. The wire K is connected with one pole of a battery. From the other pole of the battery a similar wire K' passes round the wire I and is connected as shown. In order to cause the hermetically sealed chamber between the diaphragms to be expanded or collapsed, to regulate at will the amplitude of movement of the turns of the coil X, a bulb is employed to place the air under regulable tension. In Fig. 2 the ends of the coils have carbon buttons mounted on metal discs. The coil is suspended only by threads. The carbon buttons are arranged to bear with an elastic pressure against the diaphragm plates.

The first thing to note is that this instrument suffers from the same defect as the original Bell telephone—the receiver also has to double as the transmitter and the user therefore has to put his or her mouth and ear alternately up to it, yet this is almost a quarter of a century after Bell's system was invented.

Secondly, the instrument itself was an iron box joined to the 'distant' receiver by an iron wire, making the whole thing stiff and heavy. What is worse is that the internal coil is suspended by threads so that the whole thing could not be hand held but would have to be bolted to a solid support and that would make it very awkward to use—putting one's ear to it and then speaking into it alternately.

Another important point is the question of expense. Imagine the cost of, say, 100 miles of iron wire with a layer of fine insulated wire close wound round it. What is more, the battery would have to supply current to the entire circuit and it hardly seems feasible to have a network of these telephones. All these factors really condemn the whole idea to the scrap heap before we have even begun to question the scientific aspects—which are dubious to say the least.

As well as the telephone, the electric light was another area of contention. In England Joseph Swan had been working on an incandescent electric light since the 1860s and by 1878 he and Edison were virtually neck-and-neck. Domestic installations of the electric light, however, were not feasible until satisfactory electrical generators had been developed. In Europe it was probably the Paris Exhibition of 1881 that led to interest in this sphere and the possibility of commercial viability stimulated development. Although the advantages of an electric light were obvious enough in theory, people in Europe were much slower to press for its adoption than in the United States. Moreover there was a tendency to 'convert' old styles of lighting to electricity rather than to design anew with a fresh start. A good example of this kind of conservatism is the battery operated electric light designed by E R Dale who was an Associate Member of the Institution of Electrical Engineers. This was a straight replacement for the common candle—and probably not as good, especially when the battery was becoming exhausted—yet it appeared as recently as 1900, almost twenty years after the Paris Exhibition.

I send you a block of a portable electric candlestick which meets a great demand. In my younger days I lived in a country house. I believe you could hardly have put your finger between the spots of candlegrease on the stair carpet.

Dale's electric candle, 1900

The candlesticks are made in various designs, the one illustrated being made of pure white hard metal, quadruple silver and gold plated, mounted on finely polished ebony, sycamore, and oak stands. To light the electric candle it is only necessary to turn switch on button, and turn it off to extinguish the light. When a new battery is required, slide out the bottom of base, take out the old one and replace by a new one.

In Britain, where gas lighting originated with the experiments of William Murdoch in 1792, there was a distinct tendency to stay with the old and familiar rather than to risk the expense of installing the new and somewhat fearful electric light. As ever, the majority simply could not afford it. Indeed in rural areas even gas was out of the question because of the need for a gas works and runs of piping. Consequently it was mostly in towns and districts within easy reach of a gas supply that gas lighting was to be found. Even after it had become cheaper to install electric lighting than gas, poorer people preferred gas—because for the same cost it provided *heat* in the winter as well as light.

A development with gas lighting that contributed to the extension of its usefulness and popularity was the incandescent mantle. One of the pioneers in this field was F H Wenham, a founder member of the Aeronautical Society. In 1894 he described some of his experiments with incandescence in the *English Mechanic* as follows:

As the intense light obtained from the heating of a pellicle of small particles of non-combustible matter is destined to act an important part in the future of economic and brilliant gas lighting, some particulars of the phenomenon may call for notice.

It may oftentimes have been observed that in the dull embers of burning charcoal from logs of oak or ash some stationary sparks appear at times, giving out an intense white light, which illuminates all surrounding objects, and lasts for several seconds, much resembling in colour and brilliance the light of the electric glow lamp.

It is a property of very small particles to attract invisible heat and disperse it in the form of light. The effect depends principally on the minute size of the radiating atoms, their non-conduction for heat, and its consequent retention, and also their absolute infusibility. A very fine platinum wire held in the out-flame of a candle gives an intense light, and has a collective power for heat by which such a high temperature is induced, that the cobweb wire becomes actually fused.

In the year 1852 some tufts of confervae, or common 'hair weed' were sent to me for examination. These were taken from some rock pools called 'Hell's Kettles' in consequence of continuous bubbles of carbonic-acid gas [carbon dioxide] rising to the surface of the water. The filaments of the weed were coated with a white deposit in which diatoms were expected to be seen under the microscope. I found that the coating was amorphous, being nothing else but a very fine deposit of carbonate of lime without structure. I held a few of the filaments in a gas-flame and was struck by the intense light given out. The vegetable substance of the filament was of course consumed but the lime coating, being infusible, retained its form. I then attached a fringe of these filaments from a ring of wire, and suspended it so as to encircle a small solid or Bunsen gas-flame, and got a fine white light arising entirely from incandescence of the lime. I took no further hint of this at the time as the least puff of air was sufficient to break up the fabric.

In the year 1880, having then in view improvements for increasing the intensity of ordinary gas light, I called to mind the foregoing experiment and reproduced it in a more substantial and stable form. I first constructed a cage of very fine platinum wire to inclose a Bunsen gas-flame; knowing that thin platinum in time became disintegrated and wasted by the continued action of gas-flame, I brushed on a protective coat consisting of a paste of very fine fluor-spar. The heat from the flame acted upon both the inner and outer surfaces of the cage. By this means I got a brilliant light from an otherwise invisible flame. The chief objection to this light was that it had a somewhat greenish tinge.

Illuminating power of gas

The fluor-spar is quite infusible and is decomposed into its elements by continued heating. The light, though very economical in gas consumption is far surpassed in this respect by the now well-known regenerative lamp, a form which I soon afterwards introduced into the market, but the 'incandescent' gas-light has the merit of great simplicity, and can be understood and managed by anyone, and consequently has recently been very popular, and improvements will probably be forthcoming that will greatly enhance its use and efficiency.

An unusual way of 'improving' the illuminating power of gas was tried out in 1867 by Thomas Fletcher of Warrington. His name later became well known both in the home and in science laboratories the length and breadth of Britain as a maker of gas burners of all kinds. His 1867 experiment, however, was aimed at reducing the cost of gas. This is how he did it.

NEW APPARATUS.
READY AND IN PREPARATION.

READY.

SAFETY BUNSENS (Fletcher's Patent). A new high-power Burner for confined spaces, small in size and of enormous power.

SMOOTHING IRONS, Self-heating (Fletcher's Patent), for fine Laundry work. Three sizes.

HIGH-SPEED BOILERS (Fletcher's Patent), for Boiling water **AT ANY REQUIRED SPEED** for Tea, Coffee, and Restaurant work.

LABORATORY BUNSENS for ordinary Chemical use.

IN PREPARATION.

A NEW ATTACHMENT FOR EXISTING GAS COOKING RANGES (Fletcher's Patent), for utilising the waste heat of the Oven in heating water for washing dishes, &c.

INSTANTANEOUS BATH HEATERS (Fletcher's Patent), for supplying **PURE** Water fit for Cooking or any purpose at any required speed.

NEW COOKING RANGES (Fletcher's Patent), for utilising the waste heat of the Oven in heating water. Five gallons of water heated 1° per minute, by the waste heat of the Oven without any cost whatever.

NEW FURNACES for Glass and China Painting.

INTERNALLY HEATED TAILORS' IRONS, which can be heated from an ordinary Gas Bracket.

THOMAS FLETCHER,
THYNNE STREET,
WARRINGTON.

London Offices and Showrooms—83, UPPER THAMES STREET.

Some months ago I got a tin box made 10 inches long, 5 inches wide and 2½ inches deep, filled with coarse sponge and having a ¾ inch hole in the top, over which a screw cap is fitted gas tight. Into the top of the box at one end, the pipe from the gas meter is soldered, and from the top at the other end issues the gas pipe which supplies the house.

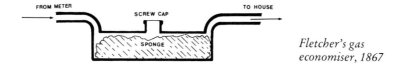

Fletcher's gas economiser, 1867

The chamber thus formed I keep about half filled with benzole, which is taken up by the gas in passing, thus making the gas of more than double the illuminating power. To put the benzole in the chamber, I turn the gas off at the meter, remove the screw cap and pour it in, replace the cap and turn the gas on again. Naphtha and several other hydrocarbons will answer the purpose, but none so well or cheaply as benzole. My average consumption of benzole is 1 quart in 3 weeks, the price being 4 shillings per gallon. My lowest average consumption of gas per week before using the above was 850 cubic feet. Since the addition has been made, the greatest consumption in one week has been 460 cubic feet, showing an absolute saving, after paying for benzole, of 10 pence per week, or about 1s. 3d., although the amount of light has been very much greater. I use smaller burners than before, a 3ft burner giving fully the light a 6ft formerly did. The light is also much whiter. No attention is required, except to add benzole about once a week, which can be done in a minute, and of course as there is a slight escape of gas I do it in daylight, so as not to require a candle. Benzole may be purchased at any druggist's, and is best kept in a tin bottle with a cork.

I use gas exclusively for cooking, burning it in the Bunsen burners, that is gas and air mixed, and find they still burn perfectly clear and blue, without the slightest smell or smoke. The economy in these, however, is comparatively small, which accounts for the amount of my saving being less than I first stated, and less than it would be if used for illumination only.

One aspect of using gas in the home that is not often given much thought is the polluting effect it has on the air. In 1900 this was investigated by a teacher at Manchester Grammar School by the name of Francis Jones. His findings were reported in the *Manchester Guardian*.

Mr Francis Jones of the Manchester Grammar School has made a valuable series of experiments to show the effects of various methods of heating and lighting on the atmosphere of rooms. The chief conclusion to be drawn from them is that gas, for heating or lighting, should be avoided by all who wish to live in reasonably pure air. It is usually estimated that air becomes harmful to health when it contains more than 13 parts of carbon dioxide in 10,000. The only combination which Mr Jones found suitable to keep an inhabited room up to this standard through an average winter day was that of a coal fire and electric light, which never raised the proportion of carbon dioxide so high as 12 parts in 10,000, the average of the experiments being made after ten o'clock in the evening, when, of course the air was at its worst. The ordinary combination of a coal fire with gas for lighting at once ran the average amount of carbon dioxide up to 27 in the evening. With a gas fire and electric light it was found that the injurious gas rose rapidly for the first two or three hours, and then remained steady at 14 to 15; if gas was also used for lighting it rose to 32. Mr Jones apparently did not experiment with oil lamps. The worst results of all were produced by the use of a gas cooking-stove; even though this was fitted with a flue leading into a chimney it raised the proportion of carbon dioxide to 40 parts in 10,000 and when the burners on top were lighted this was more than doubled, averaging 84, or more than six times the limit of safety.

If Mr Jones' results are accepted at face value, then the lesson is clear—don't have gas lighting and above all don't use a gas cooker unless you are prepared to risk suffocation from carbon dioxide pollution. Despite his awful warning, the gas industry has managed to survive, and indeed continues to thrive on sales of cookers and other appliances. There are probably more gas fires—and gas-fired central heating systems—in Britain now than ever before, and certainly far more than open coal fires. One use of gas, however, that would more than likely have thrown Mr Jones into fits, was a gas-fired alarm clock. This lethal device was the invention of Thomas H Purves in 1875. It reportedly cost 1s 6d [7½p] and was claimed to be capable of waking the soundest sleeper. Not only that it brewed a pot of coffee and blew a whistle at the same time.

The device required an ordinary clock suitably modified to turn up a gas light, which of course had to be left on all night, but turned down very low—itself a grave risk. If the light accidentally went out, then the sleeper would be asphyxiated! However, for anyone wishing to try the system, this is what you have to do:

Fasten a knee-pin A, ⅛ brass, say, at the particular hour you wish to rise, taking care it does not touch the minute hand. On this pin put a cord with two loops, the under one on the pin, the upper one loose to meet the hour hand coming down. This hand must be sharply bent at the point to catch the cord D. It immediately sets away with it, the pull opens the cock, which must be greased and left as slack as safety will permit. The gas (which has been left just on all night) is turned fully on, then the coffee boils, making your breakfast ready before it wakens you; then the steam that is generated blows the whistle B for fully an hour, and informs you that your morning cup is ready. If this don't do, I am afraid that you will have to engage a policeman.

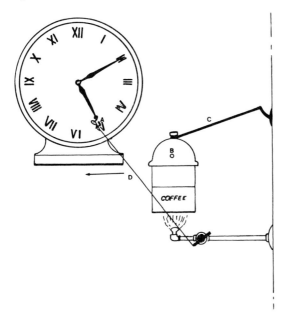

Purves' gas-powered alarm, 1875

Could this have been the origin of the well-known line in the music-hall song: 'If you want to know the time, ask a policeman'?

What Mr Purves has overlooked is the inevitability of the whole affair stopping the clock anyway. If the hour hand catches the top loop in the string as intended, but fails to lift the other loop off the 'knee-pin', then it would stop immediately. Even if that part worked satisfactorily, as the gas tap lined up with the string due to the pull on it, that too would stop the clock, although that would not really matter if it had succeeded in turning up the gas without extinguishing the flame. Presumably the 'victim' jumps up and detaches the string before the neighbours start banging on the wall because of the steam whistle going full blast! An interesting question is, would a gas tap, loose enough to be pulled round by the hour hand of an ordinary clock, still be sufficiently gas tight to prevent a lethal leakage? If the tap did leak, there could be a spectacular bang at breakfast time.

Speaking of neighbours and 'noises from next-door' a regular reader of the *English Mechanic* appealed for help from his fellow readers to solve the problem of his next-door neighbour's piano:

I am troubled with my next-door neighbour's piano. I should be greatly obliged if some of your readers could inform me what is the best method to absorb or keep back the sound. My drawing room is separated from my neighbour's by a 14 inch wall.

Two correspondents responded with advice as follows:

First suggestion:

I would recommend putting a wall 6 in. from the present one, not letting the scantlings touch the neighbour's wall, but fasten them to the floor; then fill the space with straw and mortar in alternate layers.

Second suggestion:

Ask the neighbour to stand his piano on thick rubber slabs and then line your side of the wall with battens and canvas. You cannot altogether deaden the sound, and if you are obliged to live in the room one can pity you, although I would not mind living next to, say, Paderewski.

It could have been worse if the noisy piano kept the baby awake. However, in that event help was at hand in the shape of an electric cradle for rocking the baby gently to sleep.

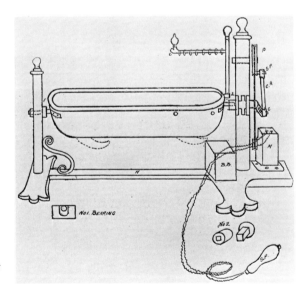

Battery-powered electric rocking cradle, 1897

In today's climate of 'energy consciousness' and pressure to build energy-saving houses and so on, it is perhaps surprising to find that energy conservation is nothing new. The Victorians were for ever trying to get something for nothing, or at any rate to find the cheapest possible way of doing something. In the days before the widespread use of small electric motors most small appliances had to be manually operated and this led many people to find an alternative power source that would run without effort or attention. Such is the case with the little motor devised in 1893 by Mr Frank Mitchell of Bouchon Works, Redman's Road, London. This little motor was intended to run items in the home that required low power, such as ornamental fountains, fans and so on. It could be operated by any convenient source of heat—including the sun.

Mitchell's heat motor, 1893

This motor consists essentially of a wheel mounted on trunnions. The wheel is hollow and divided into a number of compartments which are filled with water or other vaporisable fluid, and sometimes charged with a volatile body. The opposite pairs of compartments are connected together, and the whole is permanently sealed up. Since no chemical change takes place, one charge is sufficient to last for years of constant work, the wheel being to all intents and purposes a solid one.

To set the motor in action, it is sufficient to expose one side of the periphery of the wheel to the sun's rays, to a feeble gas jet, or even to the heat evolved by the human hand held near it. No condenser or any other device for concentrating the sun's rays is needed, and provided the heat be kept constant, no governor is required to insure regular speed. The application of heat on the one side causes a variation in the pressure of the fluid or vapour in the chambers, and this, by upsetting the equilibrium of the wheel, causes it to rotate with considerable force, which is proportionate to the difference between the heat applied and the normal temperature. One great advantage in this motor is the absence of all risk, and this conjoined to the fact that it runs perfectly silently and without dirt, is a grand recommendation. A small motor standing but a few inches in height, and actuated by the heat from a common gas burner flame turned down to the size of a small pea will work a small fan or fountain &c. I present your readers with an illustration of this novel device which, in point of durability, promises to be the most durable motor extant.

By this invention Mr Mitchell has at last solved the problem of obtaining power direct from the sun. The apparatus when arranged as a sun motor requires no attention. The enormous advantages this will offer for ventilating and other purposes in countries where the sun's heat is so intense will be obvious.

A much better method of using solar energy was thought up by a Mr Burton of Indianapolis in 1898. The *Indianapolis News* carried a report of how Mr Burton had used solar energy to wind up his clock.

In this invention the axiom of heat expanding and cold contracting is the basis. The clock is wound by changes in temperature, the principal force being in the day and night differences. Mr Burton found there is an average difference of 20 degrees in the temperature of the night and the day. The day of course is the warmer. The heat of the day expands the atmosphere, and the lower temperature of the night contracts it. This is how Mr Burton applied the force to his clock—an ordinary old-style clock—using a weight:— Outside of his house he has a tin tank, 10ft. high and 9in. diameter. It is air-tight. From it a tube runs into the cellar. This tube leads to a cylindrical reservoir which receives the air from the tank. In this reservoir there is a piston, whose rod moves with a ratchet between the chain on which the ratchet depends. The heat of the sun expands the atmosphere in the exterior tank, thus forcing any excess into the reservoir near the clock. During expansion the piston rises. In the night-time the contraction of the air in the exterior tank reduces the air in the reservoir and the piston lowers itself. The ratchet arrangement winds the clock.

Although the explanation is clear enough, it is obvious the newspaper reporter was no physicist. As every school-child knows, gases always expand to fill the container whatever the temperature and so it would have been more correct to have said that *pressure* changes brought about by the temperature variation moved the piston.

Without the convenience of central heating, obtaining enough hot water for a bath could be something of a chore. A correspondent to the *English Mechanic* in 1887 described an ingenious device invented by the French that he had come across many years previously.

I think that 30 years have elapsed since this extraordinary article was suddenly brought to my notice. I had ordered a hot bath at a French hotel; when

I entered the bathing-room I only found a fine large receptacle of clear cold water. I rang the bell and told the waiter I wanted a *hot* immersion, not a cold one. He replied the fire was not quite ready for use. This statement made me imagine that I should have to wait long for the boiler to do its work, but was startled a few minutes later on by the apparition of a curious stove brought in and placed in the water of the bath. It was half-full of incandescent charcoal glowing delightfully by aid of the air blast rushing down the side pipe B.

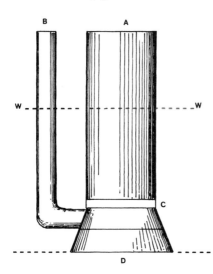

A French bath heater, 1857

The stove A was made of light sheet iron, riveted and brazed, with a heavy bottom at D; a perforated plate inside at C supports the live coals and helps the air pipe B to act. By placing an inverted metal pot under the stove it can be raised higher so as to warm a larger volume of water if necessary; but this retards the process, as might be expected. Wherever charcoal can be procured, as in Canada and India, this apparatus is of value. The fuel must be *smokeless*.

Eleven years later another correspondent was writing on a similar theme with surprisingly little advance in the technique—except that in this case the person taking a bath ran a severe risk of being squashed flat under a second bath full of water suspended from the bathroom ceiling.

... I had a small circulating coal stove attached to the end of my bath, and on the same level in the ordinary way; but it took so long to heat up the bath in this way—it had to be run off, refilled with entirely cold water, and again heated up if a second bath was required—that I found it quite impractical. I then rigged up a cistern, consisting of an old bath, close under the ceiling, about 6ft above the floor-line of the bathroom, connecting ½ inch main water pipe to it by a ball and swimmer-cock for closing off automatically when full. The stove previously connected to the bath I connected up to the cistern, and by so fixing I was able to provide a store of heated water from which to draw as required. I was thus enabled to get a couple or three baths successively without too much loss of time. Fire lit at 12.25, temperature of room 48°; temperature of water 41°; first bath taken in 105 minutes, second bath in 190 minutes, and third in 255 minutes from lighting up. Temperature of bathwater 91°, and quantity used in each case 5.2 cubic feet, increased as required by adding cold water. Temperature of room now raised to 63°. Stove heated by coal, and having a fairly good draught ....

This anonymous writer had by a goodly number of years anticipated the modern system of having a heated storage tank. It must have been quite a luxury to have a marvellously heated bathroom, but what a job keeping the boiler supplied with coal! It must have been a fairly substantial ceiling too, to have in excess of 150 kg of hot water suspended from it in a bath that must have weighed a fair bit empty. What happened if the ball valve stuck open? If the bathroom windows steamed up, the *English Mechanic* recommended a saucer of sulphuric acid placed at the bottom of the window, so as well as the risk of being drowned, boiled, or flattened, the intrepid bather took a chance on being dissolved too. This particular bathroom sounds like an ideal setting for an Agatha Christie 'whodunnit'.

If after having succeeded in having a bath (or in this particular example, three baths) the lady or gentleman of the house fancied a refreshing cup of tea, it might have been provided in a 'Universal Teapot' supplied by Messrs George Donald & Co of Central Chambers, Glasgow. This was available in 1896 and contained what can only be described as the forerunner of the tea bag. The only difference was that the 'bag' was a stoneware vessel which contained the tea leaves.

The idea can, of course, be applied to any stoneware or china pot. The object aimed at is, of course, the separation of the spent tea-leaves from the hot infusion, and this Messrs George Donald & Co accomplish by attaching to a chain which passes through the lid of the pot a small vessel for containing the tea (see illustration). This vessel or infuser is allowed to rest at the bottom of the pot until the tea is 'drawn' and is then elevated into the lid by means of the chain and kept in position by a pin passing through one of the links. The 'Universal' teapot is made in brownware, stoneware and china, the best materials of which teapots can be made; and for the end in view—namely the making of an infusion and not a decoction—the contrivance is all that can be wished.

It is well to remember that the kettle should thoroughly boil when the water is used, and that three to eight minutes be allowed for the 'draw'.

Unboiled water is simply useless for the proper extraction of the flavour. The stoneware pot can stand any ordinary heat, so that tea can be kept hot any length of time without a trace of bitterness. It can even be reheated with the same result. As to economy, it is claimed for this system that for ten ordinary cups (No 2 pot) *two* teaspoonfuls of the dry leaves will give a better flavoured beverage than *three* will in an ordinary pot. As soon as the tea absorbs the water it sinks and becomes so adhesive there is not the circulation afforded by the infuser which is in and surrounded by the hottest water. These pots may be procured at crockery, grocery and hardmongery stores for 1s 6d, 1s 9d, and by parcel post 2s 6d each in Rockingham ware of neat design.

The 'Universal Teapot', 1896

A fundamental addition to the bathroom was of course the 'water closet'. A development was planned by a Mr J G Jennings in 1868 with the intention of extracting the waste for use as fertiliser, whilst still retaining the hygienic water trap:

> ... whilst the excrementitious matters which the basin may receive, together with the small quantity of water which the basin holds are discharged into a receiver so as to be applicable for use as manure, the comparatively large quantities of water which are used to scour the basin are kept out of this reservoir and directed away into a sewer....

Unfortunately no illustration of the device was given and it is not known whether the idea ever became a reality. It might well have been frowned on as being 'not quite nice' to grow one's vegetables for the table on the effluent from 'the smallest room'.

Turnip bread

Food and diet were given close attention by the more affluent members of society, but the poor had to settle for what they could get. Prices of staple foods—especially bread, meat and potatoes—rose quite dramatically at times, while wages failed to keep pace. Those who were unfortunate enough to have to depend on the workhouse had a very thin time of it, being forced to subsist on a diet largely composed of bread, porridge and pea soup, with little or no meat. All kinds of dodges were tried to make cheap substitutes for the 'real thing', a notable one, for example, being 'turnip bread'.

At the time I tried this method, bread was very dear, insomuch that the poor people in the country could hardly afford themselves half a meal a day. This put me upon considering whether some cheaper method might not be found than making it of wheatmeal. Turnips were at that time very plentiful. I had a number of them pulled, washed clean, pared and boiled. When they were become soft enough to mash, I had the greatest part of the water pressed out of them, and afterwards had them mixed with an equal quantity, in weight, of coarse wheatmeal; the dough was then made in the usual manner with yeast or balm, salt, water &c. It rose very well in the trough, and after being well kneaded, was formed into loaves and put into the oven to be baked. I had at the same time other bread made with common meal in the ordinary way. I baked my turnip bread rather longer than the other.

When they were drawn from the oven, I caused a loaf of each sort to be cut, and found on examination, the turnip bread was sweeter than the other, to the full as light and as white, but had a little taste, though no ways disagreeable, of the turnip. Twelve hours afterwards I tasted my turnip bread again, when I found the taste of the turnip in it scarcely perceivable, and the smell quite gone off. On examining it when it had been baked 24 hours, had I not known there were turnips in its composition, I should not have imagined it. It had, it is true, a peculiar sweetish taste, but by no means disagreeable; on the contrary I rather preferred it to the bread made of wheatmeal alone.

After it had been baked 48 hours it underwent another examination, when it appeared to me rather superior to the other. It eat fresher and moister and had not at all abated in its good qualities; to be short it was still very good after a week, and as far as I could see, kept as well as the bread made of common wheatmeal. In my trials of this bread by the taste, I was not satisfied with eating it by itself; I had some of it spread with butter; I tasted it with cheese; I eat of it toasted and buttered; and finally in boiled milk and in soup. In all these forms it was palatable and good. When I had thus far succeeded, I had some more of it made in the same manner, and after it was baked and cold, I sent for some of my poor neighbours, giving them some of it to eat. They said there was something particular in the taste of it, but could not tell what to resemble it to. They allowed it was not disagreeable, yet, when I told them in what manner it was made, they declined eating any more of it, alleging it was not what they were used to and no persuasions were powerful enough to induce them, though wheat was then at a very high price, to make some of it for their family use. I am very much inclined to think that very good bread might, in the same manner, be made in times of scarcity with carrots, parsnips, potatoes, Jerusalem artichokes, and many other articles which may be raised at a trifling expense; the carrot puddings and potatoe puddings, which are both frequently seen at the tables of the great, have no particular taste of the respective roots they are made of; and this would, I dare say be the case with bread. It is for the interest of the community that the food of the poor should be as various as possible. Whilst their chief food is bread made of wheatmeal only, every time the crop of wheat fails, they are driven to the greatest distress; whereas, had they other ready and cheap resources this would never be the case. When wheat is dear, turnips or potatoes are frequently to be had at a reasonable rate; but if prejudice steps forward and forbids the use of them, of what avail is it?

In the event of London being beseiged by the enemy, as frequently predicted by the military and journalistic Jeremiahs of the day, the experiment of feeding our starving troops on turnip bread might ... be sufficient to drive them over to the enemy in disgust, or at any rate cause them to bombard the commander-in-chief to death with turnips.

That article appeared in *The Miller* in the summer of 1896, but it was based on an extract from an article that had been published in *Museum Rusticum et Commerciale* on 27 September 1763. It is significant that the 'experimental sample population' only rejected the turnip bread *after* they had been told what it was! The *English Mechanic* two years earlier (1894) carried a report about extensive adulteration of all kinds of foodstuffs that was then prevalent in the United States and elsewhere.

Liquorice drops are usually made out of candy-factory sweepings. Wine is frequently nothing but water with a percentage of crude alcohol from grain or the refuse of beet refineries, coloured with burnt sugar, flavoured with oil of cognac and given an agreeable woody taste with a little catechu. When one buys tea for one dollar a pound, one is very likely to pay in reality two dollars a pound, because one-half the quantity is currant leaves. Grated horseradish is sometimes composed of turnip. Flour is frequently weighted with soapstone. Sweetened water, sharpened with citric and tartaric acids, and flavoured with the oil of orange skin, makes orange cider ...

Leading the campaign for real bread was a recipe in the *English Mechanic* in 1888 which explained how to successfully bake delicious wholemeal bread. For anyone interested to try it, here it is.

Wholemeal flour requires more yeast and also more water than does white flour. To 7 pounds of whole-meal we allow $1\frac{1}{2}$ ounces of German yeast and ½ ounce of salt. Break up the yeast in small pieces in a basin, throw over it a teaspoonful of moist sugar, add a little hot water and smooth it well with a spoon against the sides of basin, squeezing out all lumps, and fill up with about half a pint of the water. At commencement have ready a large jug of hot water—be sure and have plenty. It must be hotter than lukewarm, but not too hot or it will kill the yeast. What the hand can bear comfortably is about right. In this dissolve the salt. Make a hole in the middle of the meal in bowl and pour in the barm; thicken into a batter with meal from sides, and sprinkle a little also on top; cover with a cloth and leave to rise for, say, 15 minutes. Now pour in gradually more hot water, and mix until you have used up all the meal. It is as well at first to keep out a basinful of meal in case of emergencies. You should now have a nice light dough. Knead well, cut up, and put straightaway into greased tins and let stand in a cool, warm place free from draughts for an hour, or until it has well risen. Place in a well heated oven (which should not be opened under twenty minutes) and bake.

In those days, of course it wasn't possible to specify at what temperature, or gas setting, to set the oven at in order to achieve the right conditions. What constitutes 'a well heated oven' will either have to be found by experiment or by recourse to a modern instruction manual.

The importance of common salt in the general diet—as well as for the treatment of specific ailments—came in for attention from time to time. The following article on the subject appeared in the *English Mechanic* in 1897.

Besides being a necessary ingredient in most kinds of cookery, an appetising addition to many articles of raw food, and the prime necessity in catching a bird, the laundress puts a trifle of salt in her starch, adds it in larger quantity to the water in which she washes ginghams and other coloured fabrics, rubs it well into the spots where oxalic acid has been applied to remove iron stain to neutralise the acid; or smooths the flat iron by rubbing it upon salt sprinkled on a bit of paper.

The housewife adds a pinch to the water in her bouquet-holder that the flowers may retain their freshness; scours the coffee or tea stains from the cups with it; has a portion put in the whitewash to make it adhere more closely to the surface where applied; obtains a good result by throwing a handful into the dull coal fire with no explosive results; or if the wood fire gets beyond her control and the chimney catch fire, a quantity of salt thrown into the stove serves as a damper to the flames; if the firebrick gives way in her cooking range, a paste of equal parts of salt and wood ashes mixed with cold water, and given a little time to harden, well supplies the loss; for cleaning any article of brass or copper, salt with vinegar or a slice of lemon is called into use, and followed by brisk polishing with a soft, dry towel.

The 'Home Doctor' applies the strong solution of salt and vinegar to the sprain; the heated salt bags, or salt mingled with hops, for the relief of severe pain; for a strong poultice beats together salt and the yolk of an egg; for inflamed eyelids or slight spots of skin poisoning uses the weak solution of salt and water; applies dry salt as a dentifrice, cleansing the teeth, and having a most salutory effect on the gums; as a dry shampoo, rubbing salt into the hair at night to be combed out in the morning leaving a clean scalp; administers salt straight for haemorrhage of the lungs or stomach; or a spoonful in a glass of cold water for nausea; for slight burns and fresh cuts, binds on the affected parts moistened salt; for neuralgia of feet or limbs, bathes those parts with the strong solution of salt in water as hot as is bearable.

The testimony of the 'good book' is that 'salt is good' and she who holds the three-fold position of housewife, nurse and laundress (as do many wives) must surely have proved this true....

Whether one accepts all these wonderful uses of salt or not, the most amazing thing about this article is the reference to the 'Home Doctor' treating haemorrhage of the lungs or stomach. That hardly seems a credible thing for a housewife, however talented, to tackle. Needless to say, housewives certainly *were* expected to be able to cope with just about anything in those households that could not afford servants. Of course, as in every other field of human endeavour, the *English Mechanic* contained all the advice anyone could possibly want to have—including, believe it or not, how to clean skulls:

First brush over carefully with soap-and-water to remove the dirt; then put them on a roof (or other place inaccessible to cats) to bleach; let them lie there for months, occasionally changing their position so that the sun and air permeates all parts freely....

An interesting selection of recipes was offered to the housewife who wanted something different in 1871 which would certainly be worth trying if you can stomach large quantities of butter and salt. There is no indication of how many people can be served:

Potatoe Soup a la Crème—Cleanse, peel, wash and slice up about twenty large-sized potatoes. Put them in a stewpan with one large onion, and one head of celery also sliced up; add four ounces of fresh butter, a little pepper, salt and grated nutmeg; set them to simmer on a slow fire, stirring them occasionally, until they are nearly dissolved into a kind of *purée*. Then add three pints of good white *consommé*, and after allowing the potatoes to boil gently by the side of a moderate fire for half an hour, pass them through the tammy, and having removed the *purée* into a soup pot, add if requisite, a little more *consommé* and set the *purée* on the fire to boil gently by the side of the stove, in order to clarify it in the usual manner. Just before sending to table add a pint of boiling cream, a pat of fresh butter, and a little pounded sugar.

How to Cook Salmon—As soon as a salmon is killed it ought to be crimped, by making incisions between the head and the tail, 2 inches wide and 1 inch deep. It should then be put in cold water (well-water if possible) for one hour, and placed in a fish kettle with as much cold water as will cover it, together with ¼ pound of salt and as much vinegar as will make the water slightly acid. As soon as the water is scalding hot (but not boiling) take it off and pour the water into a pan, and put it away in a cold place, leaving the fish in the strainer. Place the strainer with the fish upon it over the pan of hot fish-water to cool together, where it should remain until next day, when the fish should be placed again in the fish kettle with the same water in which it was scalded; and when it is again warmed, it is done. It must not boil. When there is more dressed salmon than can be eaten, it is particularly good fried in batter. It should be slightly sprinkled with salt before the batter is added, and if there is any Granville sauce, a little of it put on the pieces of salmon under the batter is a great improvement.

Roast Lamb—A fore-quarter of lamb, roasted, is a dish not to be despised; especially if, having carefully kneaded a good sized piece of best fresh butter with parsley, green onions and fine herbs, chopped up and mixed together, and having formed a ball of the compound, you introduce it carefully under the smoking hot shoulder before it is carved. There results from this treatment a relish and an unctuosity of which the meat is naturally deficient. If this seasoning is prepared by the dainty, cool hand of a young and pretty woman, so much the better....

If a housewife in Britain had a fancy to try some Australian or New Zealand butter in her recipes, she would have had difficulty getting it in edible condition in the days before refrigerated container ships. However, experiments were carried out in the 1890s into ways of transporting such perishable goods in an economic way. A new method was reported in the *Australasian* in 1896.

A method of carrying butter is that of packing it in a box made of six sheets of ordinary glass, all the edges being covered over with gummed paper. The glass box is enveloped in a layer of plaster of Paris, ¼ inch thick, and this is covered with specially prepared paper. The plaster being a bad conductor of heat, the temperature inside the hermetically sealed receptacle remains constant, being unaffected by external changes. The cost of packing is about one penny per pound. Butter packed in the way described at Melbourne has been sent across the sea to South Africa, and when the case was opened at Kimberley, 700 miles from Cape Town, the butter was found to be as sound as when it left the factory in Victoria. Cases are now made to hold as much as 2cwt of butter, and forty hands, mostly boys and girls, are occupied in making the glass receptacles and covering them with plaster. The top, or lid, however, is put on by a simple mechanical arrangement, and removed by the purchaser equally easily...

In view of the predominantly wet weather experienced in the British Isles, it is not surprising that the question of protective clothing came up from time to time, especially in relation to methods of making outer garments waterproof. An unorthodox method of waterproofing was described in 1882 by a John Alex Ollard, FRMS, of Enfield.

I have had a great coat, a thin serge suit, and a very light Inverness waterproofed, the mode being, I understand, that the cloth is dipped in acid, and whenever it comes into contact with moisture it closes the fibres. The advantages are the air passes in and out, and you have not the unpleasantness of a mackintosh. The disadvantages are that in a cold dry wind the clothes are cold; in moist, rainy warm days intensely hot. The wet does not run off but hangs on. With regard to the great coat, it is made of a furry cheviot, has been in constant wear for over five years in all weathers (no umbrella) and only once let the wet in when I accidentally tried the experiment of sitting in a puddle. I have carefully avoided further experiments of a like nature. You can see through the cloth. It is extremely light, but takes some time to dry. The serge suit was a failure, being too thin. It stood a heavy shower for some time, the wet congregated, and the result was a shower bath. I had no further reliance in it. The Inverness is now being worn, and answers fairly well. Cloth when thus waterproofed will hold a basin of water provided nothing touches it so as to cause suction, which shoulders and knees generally do. There is a smooth brown (also black) cloth which I would recommend, but I believe it is expensive, at least at my tailor's, who dwells in the West-end. I think waterproofing adds about 5 shillings to the cost of a coat, and is done before the coat is made.

In 1898 a housewife named Mrs A S Hunter sent in an enquiry to the *English Mechanic* about the relative merits of wool and linen for clothing. Two responses came from readers, one virtually cancelling out the other:

In reply to A S Hunter's enquiry as to the comparative advantages of clothing of linen and wool, the fact that wool has been adapted by evolution to serve as a covering for animals which, in many physiological respects are analogous to ourselves, is strong presumption in its favour. The structure of wool, as seen under the microscope, confirms this theory, and in practice wool is the most comfortable and protective covering material. Cricketers and athletes generally have from time immemorial chosen woolen clothing for violent exercise, and what is good under such conditions is suitable when the body is at rest, as transpiration never ceases, although it may differ in degrees. Care should be taken that the woolen underclothing be porous, and that it be worn under outer clothing which is also of porous wool. I speak from 17 years' experience.

Lewis R S Tomalin

In reply to Mrs Hunter respecting woolen clothing, may I say that my experience is against it. I have always found that it does not keep me warm in winter, but feels uncomfortably stuffy; that if there is persperation it becomes cold and wretched in winter or summer, and that it produces excessive irritation of the skin, and this last is not because it is harsh, as those who are able to wear it imagine, but because it is hairy. I have heard a great number of persons describe these objections also, and many have lamented that they had been induced by doctors to take to it, for in this changeable climate they were afraid to leave it off, yet they described how greatly they suffered from wearing it. They were overburdened and overheated in summer, and took cold from the chill damp of the persperation.

For myself, nothing can exceed the comfortable feeling of calico next the skin, and though I have always felt the cold, I am certainly not colder in it than when I have worn either woolen or flannel. Swansdown worn over the calico keeps a person warm. Linen I have never tried, so I can give no opinion about it.

Bran.

According to a Mr J Poulsen, one household job which required no clothes to be worn at all was revarnishing the furniture. This piece of valuable advice was given in 1883:

If they [the chairs] have been varnished, well cleanse them with soda and water; no soap. After you have cleansed them and they are perfectly dry, get some Medina black, dissolve it in methylated spirit and brush your chairs over with the liquid stain; let them dry, and then obtain some coachmakers's black japan; but mind you get black japan, and not Brunswick black, as very often that is sold for black japan. When you have got it, lay it on with a flat camel's hairbrush in a hot room. Now here is the secret. You must well dust the ceiling, the walls and the floor of the room, and also see to your clothes that not a particle of dust flies about or settles on your work. You ought to be almost naked while you are doing a job of that kind, as the least particle of dust will spoil the whole of your work. The black japan, if it is good, ought to take about 40 to 50 hours drying. After they are finished, to make them look very handsome, you might pick and line them with gold bronze . . . but if you require to know any more, advertise your address.

Another important household chore—even today—is cleaning shoes. Even before cleaning them, great care should be made in *buying* shoes, and what is more they should only be bought in the afternoon according to an article in the *English Mechanic* in 1897.

Buyers should never go in the early morning to buy boots and shoes. If it is remembered that activity and standing enlarge the feet, and at the latter part of the day they are at their maximum size, there would not be so many complaints of shoes being tight, which at the time of fitting seemed perfectly comfortable.

Shoes, like gloves, wear longer and better if kept for some time before using; and it is wise to keep several pairs for a week before wearing them, and several pairs to alternate with. Never wear a shoe too small or that does not fit when you first put it on, for misery more complete than a shoe that pinches does not exist.

A shoe should be washed every now and then with a wet rag and oiled overnight. In this case a fresh application of blacking restores the brilliancy of the leather. A wet shoe must never be placed too near the fire, for it will become hard and stiff. The way to save a shoe that is wet from an early grave is to wipe it off and then apply an oil or a cream by means of a soft piece of flannel or cloth. Wear old shoes in bad weather.

Patent leather should never be handled until warmed, and they can be made smooth and bright by cream rubbed in by a cloth or by the palm of the hand which is better.

If shoes are washed once a month, they will be soft and impervious to water.

Those who suffer from aching feet should occasionally sponge the insides of their shoes with a moderately strong solution of ammonia. The shoes must be perfectly dry before they are put on.

The way to clean kid boots, which will not bear blacking, is to roll a strip of flannel, 4 inches wide and a good yard long, into a wad and sew it tightly. Dip it into a saucer filled with a few drops of olive oil and good black ink. Daub the shoe all over, and, taking a fresh flannel, rub the shoe until it is dry. By this means the painful approach of purple and the dreaded white cracks will be delayed.

A fine polish, and one that will make the leather last longer than the ordinary blacking does, will be obtained if the following mixture is used:— Two ounces of ivory black, three ounces of treacle, and one pint of vinegar. Mix them together, and, having also stirred five grains of sperm oil and six drachms of oil of vitriol [sulphuric acid] work all the ingredients together.

Tan shoes should be washed once a week with saddle soap before applying polish which can be made by mixing one ounce of muriatic [hydrochloric] acid, half an ounce of alum, half an ounce of gum arabic, and half an ounce of spirit of lavender into one half pint of sour milk. Apply with a flannel and polish with a piece of fresh flannel.

The only comment I can make in relation to that advice is that I know I wouldn't survive two seconds if my wife found me walking through the house with shoes plastered with treacle, vinegar, and

Furniture polish

sulphuric acid—let alone sour milk, no matter what the scientific evidence of its benefit for shoe leather. Furniture polish offers much safer territory, and indeed those who own fine pieces of antique furniture made of quality woods might appreciate having an 'old-fashioned' polish to hand to keep them in good condition. Two slightly different varieties are offered:

Furniture polish 1: Beeswax, 1 ounce; white wax, ¼ ounce; Castile soap, 1 ounce; all to be shred very fine, one pint of boiling water poured on, when cold add turpentine ½ pint, spirits of wine ½ pint. Mix well together; rub well in with one cloth, and polish with another.

Furniture polish 2: Place 4 ounces of beeswax in a tin, and immerse the latter in a vessel of boiling water standing over a closed fire. When the wax is thoroughly melted, remove from the fire and add a tablespoonful of turpentine; again warm the mixture and keep it constantly stirred, add more turpentine as required until the desired consistency is obtained. Great care is needed to prevent ignition of the vapour given off. The polish should be used when slightly warm.

In the latter case I can speak with some conviction having tried it and found it first class. The warming *must* be done very carefully and *only* by standing the tin holding the wax in a large saucepan of water. The turpentine (the genuine article must be used, not turps substitute) must be added no more than a spoonful at a time and the mixture kept well stirred. A good tin with a tight lid is required to keep the polish in and to prevent it becoming hard.

Anyone who is keen on do-it-yourself plumbing could have the same accident as an anonymous correspondent to the *English Mechanic* in 1875 who had the misfortune to drop molten lead on the best carpet and who appealed to fellow readers for help before his wife found out. Three methods of removing the lead were offered, each one worse than the next, and more than likely to do more harm than good.

Removing Molten Lead from Carpet

Method 1: Pour a little mercury on the lead and get off the crystalline paste so formed with a stiff brush.

Method 2: Get a large ladle full of plumber's solder and make it hot enough to light a piece of dry paper, put a little grease from a composite candle on the lead to form a flux, dip in the molten solder and the lead will come off. Do not keep it in long enough to burn the carpet.

Method 3: Take up the carpet and place the parts where the solder is over a bucket with water in to steady it and cool the metal. Then with a plumber's hot iron (not a copper bit) and a stout piece of tin, hold it close to the metal which will drop through into the bucket.

If the carpet had been a white goatskin rug, on the other hand, all might not have been lost.

There are two methods of cleansing white goatskin rugs. If not very much soiled, wet a soft cloth with naphtha and rub the hair vigorously, doing a small portion at a time, then hang the rug upon the line in the open air that the odour may disappear. Do this work in the daylight and have no fire in the room while using the naphtha. If it is necessary to wash the rug, choose a cool windy day for the purpose. Throw a half-pint of household ammonia into a tub containing about four gallons of water. Place the rug in the tub and allow it to remain there thirty minutes; shake thoroughly in the water, rinse carefully in luke-warm water and hang in the shade in the open air. When dry it will be found very stiff, but may be softened by hard rubbing and combing with the fingers.

Dangerous chemicals weren't only used for cleaning shoes, silverware or carpets. They were also to be found in cosmetics in the nineteenth century. Many such preparations are now known to be carcinogenic. Here is an example from 1897 advocating the use of benzoin—a solution of phenyl benzoyl carbinol—in antiseptic ointments.

As a valuable astringent for tightening and improving the skin, benzoin lotion is unequalled. It also contracts large pores and prevents incipient wrinkles. Simple tincture of benzoin, which in appearance resembles pale sherry should be used, and although it is occasionally applied in rose-water, yet in this form it is too harsh and drying to the skin. We should advise our readers to purchase the tincture and mix at home—as all chemists do not keep the simple tincture, and use Friar's balsam, which contains aloes and other drugs, as well as benzoin, and also tends to darken the skin. When cucumber is available, use the fresh juice, otherwise substitute the same quantity of cucumber emulsion or cream of cucumbers. Take a ten-ounce bottle and put in this the juice of one cucumber, which usually yields from two to three tablespoonfuls, according to size. Half fill the bottle with elder-flower water and add two tablespoonfuls of eau-de-Cologne. Shake well, and then add slowly half-an-ounce of simple tincture of benzoin, shaking it now and then. Fill up the bottle with elder-flower water, and the lotion is ready for use and will keep a twelve-month.

Another method is to spray the face with cold water for a quarter of an hour before going to bed. This process will be found especially beneficial during the winter months, when the long evenings are usually passed in warm rooms heated to an excess by fire and gas, than which nothing is more destructive to the complexion.

A rather more 'heroic' treatment for the complexion was offered in 1867 and was advertised as being 'quite as efficatious applied to a machine as it would be to the complexion of the machinist'.

Pulverised egg shells, one pound; calcined magnesia, 2 ounces; terra alba 1 pound; marsh mallow root, 1 pound; orris root, 1 pound; gum benzoin ½ ounce; rose-water, 1 pint; otto of rose, 5 drops; cascarella root, ¼ ounce. Beautiful for ever—removing all impurities from the complexion.

A task that very probably would have been the responsibility of the housewife was worming the cat. Here is what to do:

First go to the druggist and buy a good areca nut, not one all cracked and worm-eaten; then ... grate off as much as [you] can pile onto a sixpence, mix it up with either butter or lard, and smear it on cat's side. As the cat is a very cleanly animal, it will lick the lard off, and of course the areca nut, thus saving you all struggling and scratching. An hour after this is done give half a teaspoonful of castor-oil. The cat must not taste food for at least 12 hours before giving the areca nut, better still if 24 hours. Repeat medicine after a week.

We have already seen that the best way to make a good cup of tea is to use a 'Universal Teapot'. However, a rather more 'scientific' way might appeal to those not possessing one. The following method was sent in to the *English Mechanic* in 1900 by an anonymous correspondent who evidently had spent many years in the Far East. He also explained how to make 'good' coffee.

For tea, fill a pan (which is enamelled white inside) with good, cold water; put it on a brisk fire, or *wickless* oil stove. As soon as it boils, throw in the tea, take it off the fire and stir vigorously in a bright light, watching the colours that come with as much care as if you were tempering a milling cutter. Have handy an exact counterpart of the first pan, and as soon as the colour comes 'just so' about a minute and three-quarters to two minutes, pour off into the second pan all the liquid, leaving the leaves behind. Put the full amount of water you will require into the pan in the first instance, and you will agree that this way is alright.

For coffee put the amount of coffee you will require for four days into a glazed earthenware jar, say, three pints, and on to it put this amount of clear cold water. Stir well and leave for twelve hours, stirring when you happen to pass. Then boil well, and when cold and quite settled, decant off clear portion to glass water bottles or other similar receptacles. Heat up to drinking point when required. Mix with its own bulk of water, for it is best to make it up to double strength for keeping quality: it will keep good for four days about ... you can make it day or night in a minute if you always keep some of the stock solution on hand.

You must of course have good tea and coffee. In the first place, get it if you can, from some planter direct. Get a lot at a time, as when you get used to one brand, you do not take to another so much, even if it is as good. This applies to coffee as well as tea.

Although this sounds like an important step on the way to 'instant coffee', that had already been tried successfully by an Edinburgh grocer named Law in the 1830s who used to sell bottled coffee extract. A tablespoonful with boiling water poured over it was said to have made a breakfast cup of 'excellent' coffee.

Home-made wine was another speciality of the housewife in Victorian times, especially in rural districts. It was thought to be not so wicked or sinful to drink that as to imbibe the products of the commercial breweries. The following two recipes date from 1896 and 1897 respectively.

Orange Wine—To make one gallon take 20 oranges, 1 lemon and $3\frac{1}{2}$ pounds of loaf sugar. Peel the oranges and lemon, and cut in small pieces, and soak with half the peel in 3 quarts of water. After soaking eight days, squeeze the oranges and lemons, and allow to soak again for two days. Strain, add 1 tablespoonful of fresh yeast and stir well together. Next place sugar in cask and pour liquid on, allow to work in cask three weeks. Cork up cask, and allow to stay six months. Bottle off, and the longer it is kept the better. This should be made in April.

Rhubarb Wine—Gather the rhubarb about the middle of August; clean and trim; cut the sticks into lengths about 3 inches; weigh up. To every 5 pounds allow 3 pounds of preserving sugar and one gallon of water. Place the rhubarb in a half brandy-cask, or any other wide vessel like a washing-tub. Now boil the water, and, while boiling pour it on the rhubarb. Now add the sugar; stir well until all is dissolved. When the temperature has fallen to

75°Fahr., place on the surface of the liquor a slice of toast, on which you have placed a tablespoonful of good yeast from the brewers. Let the wine ferment fifteen days. Now with a clean bowl take up as much of the clear liquor as possible, and cask it into a clean beer or other cask (not musty). Let it stand therein until December, first lightly bunged down, then more tightly. It will be fit to drink from the cask by Christmas; but if desired effervescent, may then be bottled, adding half a teaspoonful of a caster sugar to each bottle before bottling. Almost indistinguishable from champagne.

Equipped with a good supply of rhubarb wine, one can imagine wealthy Victorian families up and down the country retiring to Papa's study after Christmas lunch to watch the Queen's Christmas Message on the 'teleanimatograph'—at least they could have done if a Mr Leonard W Hall from Vale View, Windermere had been able to get any backers for his invention of a means of transmitting moving pictures by telegraph which he brought to the notice of the world in 1900. Unfortunately at the time there seemed to be very little future in it. In any case he had already been anticipated by Paul Nipkow, a German engineer who had been granted a patent in 1884 for an 'Elektrisches Telescop' (literally translated as an electric telescope) which utilised the photoelectric properties of selenium. It was Nipkow's system that was reinvented and developed into a successful electro-mechanical television set in 1924 by John Logie Baird. Nipkow is now probably best remembered for his optical scanning device—the Nipkow disc—which was fundamental to the Baird 'Televisor'.

CLEANING RUBBER SHOES

The easiest way to clean rubber shoes of any kind is to rub them with vaseline. They then clean much better and last longer than if they are washed with water.

English Mechanic 1891

PRESERVING GREEN PEAS

Preserve green peas by placing them in a bottle and sealing up, after which they should be buried in the ground.

English Mechanic 1879

Chapter 3
Philosophical Amusements, Pastimes and Hobbies

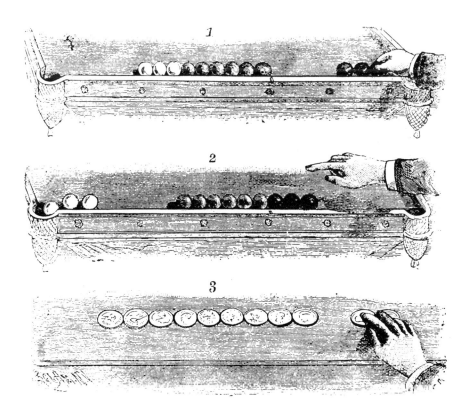

DESPITE working incredibly long hours, all classes of people in Victorian Britain seemed to find at least *some* time, however limited, for 'recreational pursuits'. At one end of the social scale such pursuits helped to while away hours of boredom; at the other end it was probably a desire to escape from drudgery that led to the creation of entertaining diversions. Of course what the very wealthy could afford was greatly different from what the rest could. Even so, amusements were not supposed to be time-wasting and frivolous. Most people seemed to want to be educated through their leisure activities and, in the same way that popular literature always seemed to present moralistic lessons, toys and games frequently had the aim of illuminating scientific or engineering principles.

If anyone was stuck for ideas as to what to do of a winter evening, the obvious place to look was in back numbers of the *English Mechanic*. An article which appeared in September 1897 under the heading 'Experiments for Entertainments' gave a whole range of scientific principles to explore at home:

The time is approaching when many people are considering what they shall do for getting up entertainments in the winter evenings. I send some notes on chemical paradoxes which may be useful or suggestive. We are accustomed, for example, to associate the idea of combustibility with paper. If it be wrapped tightly round a metallic rod, it can be held in a gas flame without burning. The metal carries the heat away from it as fast as applied, becoming hot itself. After a while it will reach a temperature ... at which the paper will burn. This same phenomenon can be more strikingly exhibited by making a vessel of paper, filling it with water and applying heat. No matter how hot the flame over which it may be, it will not burn. The water will boil and the heat be absorbed or rendered latent in the production of steam. An egg can thus be boiled in a paper saucepan.

It seems paradoxical to see a genuine metal melt in boiling water. It is a general rule that alloys melt at a lower temperature than any of their components. By making an alloy of cadmium, bismuth, lead and tin, in proper proportions, we form a compound that will melt far below the boiling point of water. A good way to exhibit this is to make teaspoons or punch ladles of it so that they will melt in the hot fluid.

Double decompositions are responsible for many of our titular experiments. By mixing solutions of ferric oxide and potassic ferrocyanide we obtain Prussian blue. The solutions may be so dilute as to be colourless. So two colourless solutions produce a coloured one, the suspended precipitate colouring the mixture. So may chrome yellow, or lead chromate and mercuric iodide, and hundreds of other reactions be made to repeat this phenomenon. The acid radicals in these cases change places with each other. By proper succession very pretty effects may be produced. Thus five colourless solutions may be made to produce a colourless, a red, a colourless, a white, and a black mixture, all that is necessary being to pour from the first vessel into the next, the second into the third, and so on.

To make two coloured solutions produce a colourless one we may avail ourselves of the power possessed by nitric acid of bleaching indigo. Two solutions of indigo are made. One contains a good quantity of sulphuric and hydrochloric acids; the other contains potassic or sodic nitrate. On pouring them together and warming, a colourless solution results, as the sulphuric acid sets free nitric acid and chlorine which destroys the indigo.

Two liquids may be made to produce a solid. This is another double decomposition. Saturated solutions of calcic chloride and potassic carbonate are poured together, when a very heavy precipitate of chalk is thrown down.

Two gases may produce a solid. This is effected by a simple combination. Ammoniacal gas and hydrochloric acid gas are both absolutely gaseous at ordinary temperature and pressure. If brought together they combine, forming a white solid substance called ammonic chloride or salammoniac. It is the substance used by tinsmiths to brighten the faces of their soldering bolts before tinning them.

The rather long article goes on and on with even more dangerous experiments—including making explosives, and worse still, boiling phosphorus with a strong solution of potassic hydrate to produce hydrogen which 'spontaneously bursts into flame'. One can imagine the devastation caused—let alone the risk to life and limb—if these experiments were carried out on the kitchen table by precocious young children, or even by adults who didn't really know what they were doing. At the very least there would have been some spectacular fireworks. One wonders too how such dangerous chemicals—nitric acid, sulphuric acid, lead chromate, phosphorus and so on—could get into the hands of ordinary people with no form of restriction. Presumably there would hardly have been any purpose in offering the suggestions if the materials to carry out the experiments had not been freely available.

The intimate relationship between electricity and magnetism was another source of amusement and speculation. In January 1867, some 35 years after Michael Faraday's pioneering researches into electromagnetism, a most ingenious device to demonstrate the phenomenon was described in the *English Mechanic*.

Electromagnetic induction

Two coils of copper wire were shown mounted respectively on top of a pair of flat corks. Underneath each cork was a strip of zinc and a strip of copper sheet, the latter being folded in such a way as to enclose the zinc strip but without touching it. The zinc strip was soldered to one end of the copper coil, and the copper strip to the other end. When the corks were floated in a large pan of dilute sulphuric acid, the copper and zinc strips formed the plates of a very effective voltaic cell which then supplied current to the coils. The two coils could be shown to behave like magnets by holding a bar magnet near each end of either coil. Moreover, depending on which way they were aligned, the floating coils would either attract or repel each other.

Apparatus for demonstrating electromagnetic induction, 1867

A hobby that was popular amongst all sections of the community, rich and poor alike, was astronomy—either by naked eye observation or by means of a telescope. It was not beyond an enthusiastic amateur to construct his own telescope for a few pence using cardboard tubes and lenses obtained from an optician. Such is the case with the following example of a simple refractor, described in *Hogg's Natural Philosophy* in 1866.

A cheap and really useful telescope may be made by obtaining a single convex lens of 4, 5, or 6 feet focus, which lens can be had of an optician for

2s 6d [12½p]. The tube may be made of paper of the required length to suit the focus of the object glass. Make also two more tubes of tin for an eyepiece, one to slide within the other, the larger one to slide in the tube of the telescope. The annexed diagram will explain what we mean.

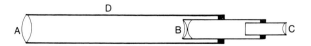

D is the body, A is the object glass, B, C is the eyepiece. This is somewhat like the Galilean telescope, but with this material difference, a compound eyepiece is used in this, whilst in the Galilean only a single eye glass is used. B is a concave, and C a plano-convex lens, each 4 inch focus. By placing B before C it doubles the power of the eye lens and gets rid almost entirely of the prismatic colours, whilst its definition approaches that of an achromatic. This eyepiece can be used as a pancratic. If you pull C out, you must push B nearer the object glass. But two or three trials will be required to adjust the glasses of this telescope which will show the satellites of Jupiter, and also the dark belt across the body of the planet.

It doesn't seem particularly safe to have an eyepiece tube made out of tin, unless there was some means of preventing the sharp edges of the metal from cutting the user's face when he or she put it up to the eye. No doubt another cardboard tube, small enough to slide in the main tube, would do just as well, provided that the sliding movement could be done smoothly enough. What is more, although 2s 6d might seem a trivial sum today, in 1866 it represented a large proportion, if not all, of a working man's weekly wages. Nevertheless, there were many who thought the sacrifice worthwhile and if need be had the ingenuity to make their own even cheaper versions of a telescope so they could pursue their hobby. Some (by no means a small number) were keen enough and venturesome enough to build their own reflecting telescopes—including grinding and silvering their own mirrors. A man by the name of Frank H Wright, for example, described in 1896 how he made himself a mirror for a Newtonian telescope entirely by hand. It took him three months, working on a corner of the kitchen table, just to grind the mirror. He did this for two hours every day *between 5 am and 7 am* before going off to do his normal 12 hours' work during the day.

To show the lengths some people went to in order to get what they needed as cheaply as possible, the following gadget is a device for moving a telescope to compensate for the rotation of the earth, and thus holds the telescope field of view steady and fixed on the object being observed. It consists simply of two planks of wood hinged together and kept apart at one end by a football bladder inflated with a bicycle pump. The bladder is fitted with a rubber tube which has a brass tap inserted at the outer end, the idea being that the tap could be regulated so as to allow the bladder to deflate at the appropriate rate when a weight is placed on the upper plank. A string attached to the upper plank passes over a pulley and is attached at the other end to the telescope barrel. As the bladder deflates, the plank gradually falls, pulling the string and hence the telescope too. If the rate of movement is exactly right, the field of view would remain steady, but if the weight was too great, or the tap was opened just a little too much, the stars would appear to travel backwards across the field of view. If the rate of falling was too slow, the telescope would appear to be left behind by the stars. In windy weather anything could happen!

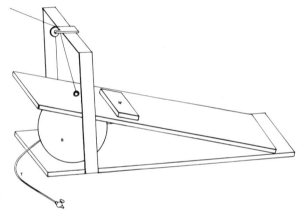

Air clock for telescopes, 1866

Amateur microscopists were not forgotten either. In 1870 a Mr Perry published a description of a neat little microscope that he had devised, although he failed to give away any secrets about the most important part—the objective lens. From the sketch that accompanied his article, it looks as though the lens was little more than a blob of melted glass stuck on a glass slide. This single lens served both as objective and eyepiece.

These microscopes are made from square brass tubes, same size as drawings; the advantages they possess are as follows:—They are made to fasten to a table or block, or better, to the box which contains them. They have a reflecting mirror, a fine screw motion for focusing, a circular diaphragm for regulating the incident light, and last but not least they are strong, cheap and well made. Fig. 1 is a front view, B is the square body of the microscope, R the revolving diaphragm with four different sized holes; L the eye lens, and S the back thumb-screw for focusing.

Fig. 2 is a section—M is the inclined mirror resting on a block of hard wood t; the light after passing through the diaphragm hole is reflected upwards in the direction of the eye lens causing the object under examination to be well illuminated; the light is regulated by using a larger or smaller hole as the case may require. The top of the microscope has a dove-tailed groove for receiving a strip of glass xx. P is the slide carrying the lens; this slide, after passing down the groove G terminates in a screw y which, passing through the thumb-piece S gives the motion for focusing. H is the hinge for fixing the box; the microscope can, however, be held between the thumb and finger if so preferred.

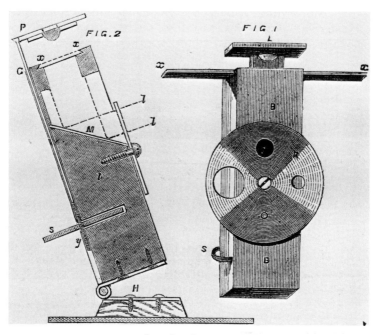

Perry's microscope, 1870

The power of this little instrument may be roughly estimated at from 20,000 to 30,000. The heart and blood of a flea are distinctly seen in motion; a few human hairs resemble a bundle of faggots, and hundreds of lively things are seen in a small drop of water....

Although this little microscope is not much more than a toy, it nevertheless would have provided a very effective introduction into the world of microscopic things, and no doubt would have given hours of absorbing amusement.

The camera too came in for its share of attention. It must be remembered that photography only became available to the general public in 1839 when Louis Daguerre in France brought out his Daguerreotype process. This was a practical improvement on the first successful photographs taken in the 1820s by another Frenchman, Nicéphore Niépce. However, it was William Henry Fox Talbot's experiments with negative pictures on paper that paved the way for still further improvements in the chemical processes of developing and fixing the image, and especially in shortening exposure times. These improvements eventually led to a simplification of the process of taking photographs and opened the door for enthusiastic amateurs to take up photography as a hobby. This kind of evolution is still going on today.

An article in *Photographic News* in 1897 described how one such amateur used bits of an old camera to make himself a new one. From the description, however, it rather seems as though the new version needed at least three hands to operate it.

First procure a large cigar box, about 9 inches by 5 inches. Find the focal distance of your lens, and cut down the box to about one inch longer than that. Strengthen the box by gluing strips of cardboard outside it, and cover the whole with black American cloth. Fix the lens in one end and knock out the other. At the open end glue cardboard strips so as to make a groove the thickness of the dark slide, and glue in it pads of black velvet so as the slide comes fairly in the middle. Now cut a little piece off the top of the box so that the whole is light tight when the slide is in.

Cover the whole of the camera with several thicknesses of American cloth. Now, by means of a piece of ground glass, focus your lens, which must be a sliding one, for all distances above, say, 20 feet; get two caps to fit the lens head and you are ready to snap shot anything. Load your dark slides; screw the lens to the focus required for the distance you are from the object. Get one cap in each hand, after pulling up the slide shutter, so that the edges of the cap in your hand touch the edge of the cap on the lens. Remove the one on the lens, and as quickly as possible replace the other on the lens head. You will be surprised if you get as good results as I have done with a camera made in this way.

A statement with which no one is likely to disagree since the whole procedure sounds like a recipe for disaster. It rather seems that the writer assumed his readers already had a thorough knowledge of cameras and their construction. For example, he makes no mention of where the slide shutter goes, or indeed how it is made, yet one is supposed to be there ready to be pulled up at the right moment. Quite what he was doing with *two* lens caps also is not entirely clear. It looks as though the procedure was to pull off one cap and thus open up the lens to the light, and as quickly as possible replace it with the second one, the time for that to happen being the exposure time. Wouldn't one cap have done just as well? Which hand was holding the camera while the other two were fiddling with the two lens caps?

The pin-hole camera on the other hand is much less bother, yet there is ample scope for experimentation even with very simple home-made versions. In 1897 Mr Alfred Watkins—a maker of exposure meters in Hereford—offered an 'exposure meter' for pin-hole cameras in the form of a table of needle sizes with which to make the pin-hole for various distances and different sizes of photographic plates.

No. of needle	Diaphragm value	Distance to plate	Size of plate
3	F/76	23	—
4	F/72	20	15 × 12
5	F/60	15	12 × 10
6	F/56	13	10 × 8
7	F/50	10	Whole
8	F/45	8	Half
9	F/38	6	5 × 4
10	F/35	5	Quarter

The Watkins' exposure meter for pin-hole cameras, 1897

According to Mr Watkins, to use this table

... note the size of plate you want, say 10 × 8. Opposite this you will find the size of needle (No. 6) with which to punch a hole in a thin sheet of brass or copper. The burr made by the needle should be cleaned off with a fine file, the needle passed through again, the hole slightly countersunk (not enlarged) and the sheet blacked. The pin-hole (the name sticks even though a needle is used to make it) is fastened in front of the camera at the distance shown in the table (15 inches for the 12 × 10 plate). No light must enter the camera except through the pin-hole. To calculate the exposure, note the 'diaphragm value' in the table, estimate how many seconds you would expose if using that diaphragm with an ordinary lens, and give the same number of *minutes* [my italics] exposure with the pin-hole.

Once again, a great deal is assumed. How a novice could estimate exposure is not explained and Mr Watkins clearly took it for granted that everyone knew what he was talking about. However, he did give an example of how to use the table, as follows:

With a 10 by 8 plate, F/56 is the diaphragm value. With a good summer light 2 seconds would be the right exposure with a lens, that stop and a rapid plate. With the pin-hole camera, two minutes would be required, and it does not hurt to give a little more.

The diaphragm value given in the table is not the actual ratio of pin-hole to distance but 60 times the value. This is done to facilitate the calculation.

The Viviscope

This explanation gives us an idea of the order of magnitude of exposure times in 1897 with an ordinary camera. Today an exposure of two seconds in bright summer light would be unthinkable, yet compared with Niépce's exposures of seven or eight hours, it must have seemed amazingly rapid. Daguerre's process in 1839 required 10 or 15 minutes exposure, although this was brought down to about 15 seconds by the mid 1840s, making portrait photography a practical proposition—provided the subject could stay frozen absolutely still for the requisite amount of time. Interestingly enough, the size of a Daguerreotype plate was used as a reference standard when it became possible to take photographs in all kinds of sizes. The reference to a 'whole plate' or to a 'half' or a 'quarter' is a measure of the size of a photographic plate as a fraction of a Daguerreotype plate which was approximately $8\frac{1}{2}$ by $6\frac{1}{2}$ inches.

By the late 1870s so much development in photographic processes had taken place that it was technically possible to obtain a rapid succession of pictures of a moving object and attempts were made to reproduce this movement by projecting the pictures in the same sequence. Even before photography was possible, there were optical toys that could simulate moving pictures, either by the use of printed hand-drawn pictures on a disc which could be rotated in a special viewer, or on strips of paper. All these relied on the persistence of vision to complete the illusion of movement.

An interesting example of a gadget for home entertainment was the Viviscope which was manufactured in the 1890s by E B Koopman of 33 Union Square, New York. An explanation of how it worked was published in the *Scientific American* in 1896.

Supported on a standard is a circular stage. Concentric with the stage a circular block about 8 inches in diameter is rotated by a hand wheel. This block is surrounded by a cylinder secured immovably to the circular stage. Attached to the disc are two wires projecting nearly radially from it, and carrying at their outer ends a block of crescent shape and which depends directly over the perimeter of the stationary cylinder. As the hand wheel is rotated this block whirls around and around the cylinder.

With the viviscope are supplied a number of endless bands of paper with coloured pictures of figures in progressive stages of movement, drawn on the zoetrope principle, the same as is followed in securing photographs for the kinetoscope and vitascope. These bands have their ends pasted together and are of such length as to fit rather loosely over the stationary cylinder and the depending block. A screen with a hole is provided which is mounted on the perimeter of the circular stage, and through this aperture the spectator is supposed to see the figures. One of the beauties of the instrument is that the screen is not really necessary, and without it the movements can be seen by an entire room full of people. When the hand wheel is turned, the block whirls around between the stationary cylinder and the endless band with the figures on it. As the block passes under each figure, by a very peculiar principle of wave motion, the figure is shifted one space forward. Thus for each rotation of the block, every figure on the band, which of course means the whole band, is shifted one space ahead, so that a perfect zoetrope effect is produced, and the figures seem endowed with life.

The Viviscope, 1896

The length of the crescent-shaped block is probably crucial to the operation of the device. It is given as one-seventh of the circumference of the stationary cylinder. The length of the paper band too will be important—or rather the amount of slack in it. The principle seems to be that a wave motion is set up in the paper band and frictional contact between the 'crests' of the wave and the whirling crescent-shaped block moves the band round in a jerky fashion such that each picture momentarily remains stationary in front of the observer before being snatched away and replaced by the next in succession. How it manages to do it one picture at a time is not clear.

A much earlier and optically more effective moving picture viewer was the Praxinoscope patented in 1877 by a Frenchman named Reynaud. In this device a paper strip containing a sequence of pictures is placed round the inside of a metal cylinder, facing inwards. At the centre is a set of plane mirrors arranged in the form of a polygon, with as many mirrors as there are pictures. On top of the polygon a candle can be placed to illuminate the pictures. The cylinder and the polygon of mirrors rotate together by means of a hand wheel and pulley.

The Praxinoscope

The principle on which the device works was described in *La Nature* in 1879 as follows:

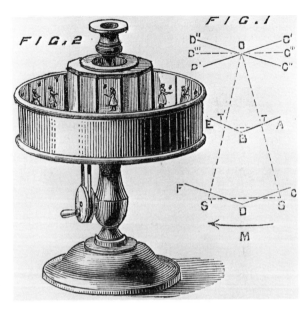

Reynaud's Praxinoscope, 1877

A mirror AB (Fig. 1) being placed at a certain distance from a picture CD the image of the latter will be reflected and visible at C'D'. When we now turn the mirror as well as the picture CD round a common centre in the same direction, so that they will occupy positions at BE and DF respectively the image of the picture will be seen at C"D". As will be seen its axis has remained unchanged. If another mirror is placed at AB and another picture at CD, the eye being placed at M, one half of the first picture will be reflected from OD" and one half of the second picture from OC'. By replacing them with another mirror and design at AB and CD the same succession of changes of position will be produced.

Mechanically the device is much simpler than the Viviscope but the main advantage of the Praxinoscope is that the image does not have to be viewed through a slot or an aperture of any kind and it is much brighter. The observer just has to look into one or other of the rotating mirrors with an uninterrupted line of vision. M Reynaud took his idea a stage further with his Praxinoscope Optical Theatre in 1892 bringing a 'night out at the pictures' a step nearer. The same principle was adopted but a long series of pictures was painted on a celluloid strip which was carried on a reel. An illustration in the *English Mechanic* (cribbed from *La Nature*) shows how it was operated.

The long celluloid strip was wound off one reel and onto another by hand and the moving image was projected onto a screen for the audience to see. Interestingly enough Reynaud used two projectors—one to present the moving pictures and a separate one to independently project a stationary image of background scenery. The illustration shows a scene from a three-character pantomime called 'Pauvre Pierrot'. It was claimed that Reynaud's 'film shows' could last up to 15 or 20 minutes, this being determined by the length of celluloid strip that could be practically handled (as well as the speed of winding the reels of course). One facility to increase the entertainment value was that the strip could be wound backwards just as easily as forwards without affecting the actual projection of the images. The obvious next step was to substitute photographic images on the strip in place of the hand painted pictures, a step taken by Edison and others and brought to an acceptable stage of perfection by R W Paul, a scientific instrument maker in London, by Birt Acres who devised a combined camera and projector, and by the Lumière brothers in France.

Reynaud's Optical Theatre, 1892

In 1896 Lumière films were demonstrated at the Royal Polytechnic Institute in London, and a series of nightly film shows was put on at the Empire Theatre which was so successful it ran for some eighteen months. By the end of the century cinemas were opening up in major towns all over the country and the age of silent movies had arrived.

Not satisfied with the thrills of seeing 'The Changing of the Guard at St James's Palace', or 'The Turn-out of the Fire Engine at Southwark', the wealthy Victorian could return home and spend the remainder of the evening listening to his tape recorder—or what he would have called his 'magneto-telephonograph'. This could be bought from the firm of Mix and Genest, Berlin who marketed the invention in 1900. The way it worked was as follows:

A wire of steel or nickel, 1 mm in diameter is wound spirally on a drum in rheostat fashion, the drum being turned by an electric motor. A tiny electromagnet with pointed iron wire cores just embracing the steel wire, slides over the wire as the drum is rotated, so that successive portions of the wire are brought into the field. Before use, the whole wire has homogeneously been magnetised crossways by connecting the coil with a battery. When receiving a message, the coil of the electromagnet forms the secondary of an induction apparatus whose primary comprises the microphone and a battery. The current in the magnet coil fluctuates in accordance with the sounds received by the microphone, and thus a varying magnetic stress is impressed on the coil wire and produces permanent poles of greater or less strength. The electromagnet is now coupled with a telephone and the drum turned in the same direction as before; the microphone currents are then reproduced by induction in the coils of the magnet, and translated into sound in the telephone. When the wire has done its duty, the magnetisations are deleted by coupling the electromagnet again with its own energising battery which restores the original homogeneous magnetisation. For longer conversations the steel wire is replaced by a reel of steel band 3 mm wide and 0.05 mm in thickness.

Even more sophisticated versions were planned, including one with an endless steel band passing over two pulleys and through a recording electromagnet, a series of transmitter electromagnets and a deleting electromagnet. This would have allowed multiple telephone outlets.

Although the basic idea of the tape recorder was there, the system was years ahead of its time. It had to be reinvented after the coming of amplifier circuits and plastic 'magnetic' tape in our own era.

Another modern development—the laser disc video system—could be said to have originated in 1892 with a device invented by a M Dumeny in France, which it was predicted would lead to the production of animated speaking photographs.

Some time ago it was announced that M. Dumeny had succeeded in taking instantaneous photographs of the lips of a speaker, and recombining them in a kind of zoetrope, so as to reproduce the original movements and enable a deaf-mute to understand what was said. He has recently improved the process and invented a new apparatus for combining the images, which he terms a phonoscope. The changes of the lips in speaking are so rapid that fifteen photographs a second are required to give a good result. Moreover, several sets of images from the same phase are taken, so that nothing essential should be omitted. The whole head and bust of the speaker is photographed, so as to get the benefit of the expression. In the phonoscope the positives are arranged around the periphery of a disc which is rapidly turned by a handle. A second disc having a window in just opposite the plates is also rotated by the same handle, but at a much higher speed than the other. A beam of sunlight illuminates the plates from behind, and the observer looking in to the apparatus sees them pass his eyes one after the other in such rapid succession as to produce the effect of a single image endowed with animation. To this end at least ten or twelve must pass the retina in a second. M. Dumeny is so far satisfied with his results that he looks forward to a time when we shall possess veritable 'speaking likenesses' of our friends. In short ... the family album will be a kind of phonoscope and contain photographs which will not only smile but speak, or appear to speak to us, as though they were alive.

On a more down to earth level, for those with no money to waste on luxuries, scientific toys could still be knocked together using discarded odds and ends around the house. The Archimedean screw pump is a typical example. The illustration appeared in *La Science en Famille* in 1896.

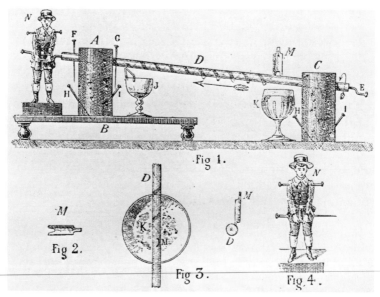

Archimedean screw pump toy, 1896

The screw itself was made from an ordinary pencil, with a spiral groove cut round it, supported in two pierced corks in such a way that one end is higher than the other. The pencil (i.e. the screw) could be made to rotate by means of a little crank at either end made from a bent pin. At the lower end of the screw an old pen nib, pressed into the pencil, was used to pick up drops of water from a full tumbler placed underneath. The drops of water picked up would then be transferred along the spiral groove until they reached a bent paper clip attached to the rim of another tumbler where they were collected. The little figure in the illustration was made out of cardboard and pins. This added a little more fun to the toy since it appeared that the cardboard boy was doing the winding. Coloured water would add to the effect.

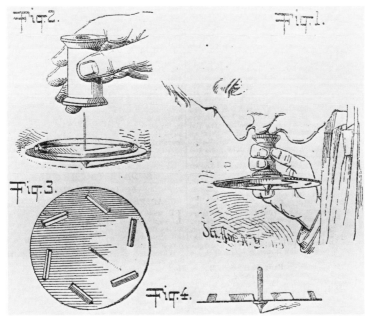

Aerodynamic top, 1892

A fascinating toy made from a cardboard disc, a pin, some sealing wax and an empty cotton reel was illustrated in the *English Mechanic* in 1892. The disc had six slots cut in it as shown in the illustration, these being actually only cut on three sides, the card being bent up along the remaining edge to form vanes. An ordinary pin through the centre of the disc was secured with sealing wax in such a way that the head of the pin protruded about 3 or 4 mm beneath the disc to form a pivot.

The illustration is fairly self-explanatory, the idea being to insert the pointed end of the pin loosely into the hole in the cotton reel and then blow down through the hole while holding a finger very lightly on the head of the pin (i.e. the pivot) to keep the disc in the blast of air. Once the disc begins to rotate it stays in position, spinning rapidly, so long as the operator continues to blow. Once the air stream stops, however, the disc will fall and if a hard flat surface such as a sheet of glass is placed underneath for it to drop onto, it will continue to spin for some time.

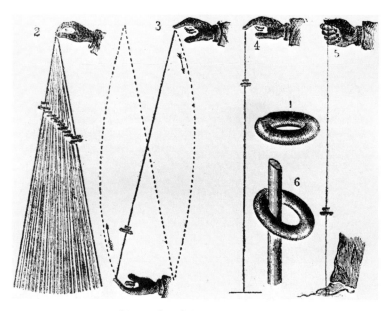

Hopkins' rod and ring experiment, 1892

Another very interesting yet simple toy is the 'rod and ring' devised by G M Hopkins and described by him in *Scientific American* in 1892. All that is required is a thick rubber ring and a smooth straight metal rod, or even a length of fine string in place of the rod. The ring is placed over the rod and twirled. As it continues to twirl it slowly falls down the rod. It even does this when the rod is swung like a pendulum (Fig. 2 in the illustration). Fig. 3 shows the rod being turned end-for-end the ring continuing to fall all the while. Fig. 5 shows the ring going through the same motions falling down a stretched string held in the hand at one end and trapped under the foot at the other. As an alternative, a metal ring could be substituted for the rubber one, and the rod replaced by a rubber tube held taut (or held straight by slipping the metal rod inside the rubber tube). The question is, how does it work?

The behaviour of the ring is due to the combined action of gravity and centrifugal force. The frictional contact between the ring and the rod prevents the ring from falling straight down the rod and it tends to spiral round by rolling on the rod because of the oblique line of contact. The principle is exactly the same as that governing the modern hoola hoop.

The boxing kangaroo is another inexpensive gadget, this time illustrating electrostatic attraction and repulsion. It dates from 1899.

The boxing kangaroo, 1899

The figure of a boxer is cut out of a card and covered on the back with tinfoil, a little larger than the figure so that it can be turned over the edges of the card. One foot of the figure is stuck into sealing wax on a small block, and to the back of this leg is secured a piece of iron wire. The figure of a boxing kangaroo in position for making an attack is cut out of tracing paper. This figure is also covered on one side with tinfoil and then is suspended by a linen thread from one end of a piece of iron wire that has a rectangular bend, the other end being set in the supporting plate so that the kangaroo shall face the boxer. To obtain the necessary electricity, we take a glass lamp chimney, stop one end of it by means of a cork and in the centre of the cork drive a nail to which is secured one end of a small piece of iron wire, the other end of the wire being connected with the wire on the back of the boxer's leg. After the lamp chimney has been carefully dried, it is rubbed with a piece of silk or fur, thus generating electricity which is transmitted to the boxer. The kangaroo is strongly attracted by the figure thus charged, which it attacks, but a discharge of electricity takes place at once and the animal is repulsed. This is followed by a series of attacks and repulses as long as the rubbing of the glass continues.

The more adventurous, equipped with a few simple tools, might decide to make an electric motor. This is a crude example powered—probably not very effectively—from a Leclanché cell and using a brass pin as a commutator.

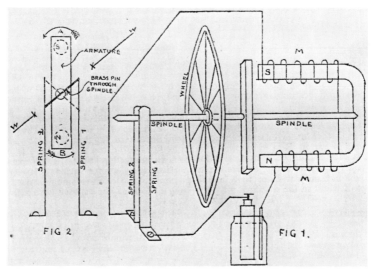

Simple electric motor, 1892

The field coil is made by winding several layers of cotton-covered wire on the straight arms of a bent piece of soft iron. The windings go in opposite directions on each arm so as to produce opposite poles at each end. The armature is a rectangular piece of soft iron mounted on a spindle so as to rotate just in front of the electromagnet's poles. The brass pin commutator is driven through the spindle at an angle to the rectangular bar and is fitted between two springy bronze strips in such a way that as the bar just aligns itself with the poles, the pin just breaks contact with one strip. The idea is that with the pin closing the circuit, the electromagnet attracts the iron bar and the inertia of the bar keeps it going when the circuit breaks until contact is made again and it receives a new impulse. The flywheel helps to smooth out the motion.

A correspondent to the *English Mechanic* in 1894 by the name of F C Allsop, the author of 'Practical Electric Bell Fitting' and similar works, offered a very pretty design for an electric motor. This had the appearance of a horizontal steam engine and worked on the same principle as the electric bell, but in slow motion. The illustration almost explains itself. Four coils wound over soft iron cores are arranged in opposing pairs so as to alternately attract a rocking iron

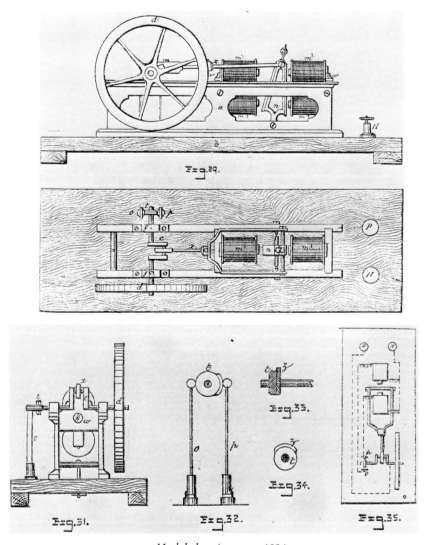

Model electric motor, 1894

armature. As it rocks to and fro the armature rotates a crankshaft by means of a long connecting rod, steam engine fashion. Current is distributed to the respective pairs of coils by a rotating contact on one end of the crankshaft, again in much the same way as an eccentric operates the valve gear on a steam engine. The coil bobbins are 1¼ inches long by about 1 inch diameter and the flywheel is 4¼ inches diameter. The side frames and other exposed parts were made of polished brass and the whole machine was mounted on a polished mahogany base board.

A delightfully simple little model which operates by thermo-electric currents was published in *La Nature* in 1895. It is in effect a wheel about 10 cm diameter made of suitable wires soldered together and balanced on a vertical needle so that it can revolve in a horizontal plane. A horseshoe magnet is held so as to straddle one part of the rim, and opposite that, the rim is gently heated by a candle flame.

A thermo-electric motor, 1895

The rim of the wheel was made of an alloy of nickel and copper, known as German silver or white bronze. The spokes were made of fine copper wire soldered to the rim. At the centre is soldered a small disc of brass with an indentation in it to receive the pivot point. The little flags shown in the illustration are not just for decoration, they can be moved about to get the wheel exactly balanced and horizontal. The finer the wire used, the faster the wheel will revolve.

A gadget with a very long ancestry—at least 2000 years—which was very popular in Victorian times was the self-acting fountain. Little ones for indoor use could be made to operate for 20 or 30 minutes or even longer, and enormous ones could be set up outside that would run without attention for hours. They all operated on the same hydrostatic principle but some were better thought out and easier to handle than others, as the illustrations show. Some were even made of glass so that the transfer of water from one vessel to another could be seen.

An outline sketch which illustrates the principle of operation of the self-acting fountain was published by a Mr A F Piggott in the *English Mechanic* in 1892. Mr Piggott described himself as a 'Gas and Electric Bell Fitter' and claimed to have made three such fountains, all of which worked satisfactorily. The procedure for starting the cycle was first to fill the upper tank with water and then to pour water down the pipe leading to the bottom chamber until it would accept no more and the water was up to the level of the jet in the centre of the basin. With any luck the fountain should then play. The air trapped in the lower chamber is compressed and passes up the connecting pipe to the upper chamber, forcing water out of the upper chamber through the jet. This water falls back into the basin to replenish that which continues to fall through the pipe connecting the basin to the lower chamber.

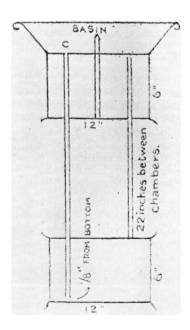

Piggott's self-acting fountain, 1892

In the Piggott fountain the jet tube actually passed through the screw-on filling cap, so that removing the cap also removed the jet which made it particularly easy to clear any particles of dirt that might obstruct the fine jet. In addition the bottom tank was fitted with a drain cock, not shown, which allowed the whole thing to be emptied.

Mr Piggott claimed that one of his fountains would run without attention for five hours, so it was of fairly ample proportions. One of these in the garden would be quite an eye-catching sight.

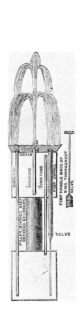
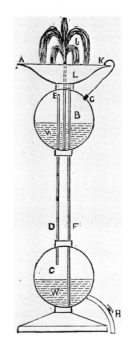
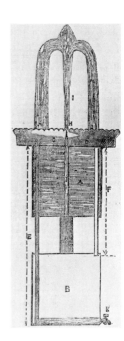

A selection of self-acting fountains

Another interesting adornment for the garden is a sundial and a particularly unusual one was described in the July 1892 issue of *Knowledge* by its inventor, Major-General J R Oliver. The unusual feature about the sundial was that it was designed to read 'clock time' rather than 'sun time'. This was accomplished by means of a specially shaped gnomon. The explanation is as follows:

Most people know that an ordinary sundial does not give clock time—sometimes the dial time is fast and sometimes it is slow compared with clock time; for sundial days are not, like ordinary days, all of equal length. When the earth is in the part of its orbit which lies nearest the sun it moves faster, and describes a greater angle about the sun in twenty-four hours than when it is in the part of its orbit farthest from the sun; consequently when the earth is in perihelion the solar day exceeds the sidereal day by more than the average amount, and the shadow of the gnomon comes round again to twelve o'clock a little later than when the earth is in aphelion.

The mean time shown by ordinary clocks is based on the division of a mean day, which corresponds to the mean or average length of the day as measured by the sundial at different parts of the year. The clock time corresponds to the dial time which would be shown by a 'fictitious' or 'mean' sun moving uniformly in the Equator at the same average rate as that of the real sun in the Ecliptic.

The 'equation of time' corresponds to the difference of time which would be shown on the dial by the real sun and the *mean* sun. It is reckoned as *plus* when the sundial is slower than the clock, and *minus* when it is faster. It is the correction which must be applied to the ordinary dial time in order to obtain mean time, and it sometimes amounts to more than 16 minutes. Although a great deal of ingenuity and thought has for centuries been expended upon the construction of sundials I am not aware that any one of the old dial makers ever succeeded in contriving a dial to show mean time. The difficulty has, however, at last been overcome.

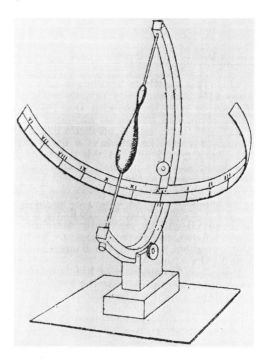

The Oliver mean-time sundial, 1892

The peculiarity of the instrument is that the time is indicated not by the shadow of a straight edge, as in the old sundials, but by the point where an equatorial circular line is cut by the edge of the shadow of a curved surface, the curvature of which is so arranged with respect to the sun's distance above or below the equator as to compensate for the 'equation of time'.

The instrument is a universal one and consists of a meridianal semicircle, the diameter of which is an axis carrying the curved gnomon and an equatorial circular arc. The latter has engraved on its concave surface a graduated line on which are marked the hours and their sub-divisions. There is a screw for clamping the meridianal arc at the proper position for any latitude, and another clamp for adjusting the equatorial arc. The dial not only indicates local mean time, but by a very simple adjustment may be set so as to show any required standard time. Thus it may be set at Plymouth to show Greenwich time. [GMT was only made 'legal time' by Act of Parliament in 1880.]

From December 25th to April 15th, when according to the almanacs the sun is after the clock, one reads the time from the following [i.e. trailing] edge of the shadow. When the equation of time has vanished and the sun is before the clock, one reads from the preceding [i.e. leading] edge of the shadow of the club-shaped bob on the gnomon at the place where its shadow cuts the hour line. Four times a year the equation of time vanishes.

The actual shape of the bob should be determinable from the 'equation of time' and data in *Whitaker's Almanac* of astronomical tables, or any nautical tables which indicate the sun's position, by means of the relationship between mean time and sun time, i.e.

Mean (clock) time = Apparent (sun) time + Equation of time.

However, to do that is not as easy as it might seem since the scale on which to work is not obvious. Regrettably Major Oliver neglected to give the actual dimensions of the gnomon that he used.

One point to note is that the 'equation of time' vanishes—i.e. becomes zero—when the sun time is the same as clock time. This occurs on 1st September, 25th December, 15th April, and 14th June each year.

An ingenious device of a totally different kind was offered to the world through the pages of the *English Mechanic* in October 1894 as the invention of a Mr R W Hill of Manchester (mentioned in Chapter 1). His invention was an 'Electrical Automatic Parrot Teacher' which anticipated by half a century a method of learning languages by listening to gramophone (phonograph) records.

Herewith are the particulars of the above apparatus which could be patented, though I am not a particularly selfish inventor, inasmuch as my little invention is not of much practical value.

The apparatus will teach automatically a parrot all languages (including the Russian, Chinese, and Sanscrit), full dialogues from Shakespeare, Schiller, Pushkin, &c.

In a cage for the parrot there is attached in a certain corner of same a small lever pivoted in the middle. One end of same is provided with platinum contacts and the other end with pincers to hold a piece of sugar. Behind this lever, a mirror or looking-glass is attached to the cage at a certain distance from the cage, not too near, otherwise the mirror may be smashed by the angry bird. A phonograph is driven by a spring motor, or a motor actuated by a weight as shown. The usual electric motor, of course, could be used, though a spring or weight motor is much simpler and will suit better for the purpose involved. The motor is provided with a round disc with one notch, a hook, and an electromagnet, as shown in the sketch. Every time the parrot snaps with his beak the sugar, the circuit of the battery is closed, the hook releases the motor for only one complete revolution of the disc and the phonograph, of course begins to talk or sing a certain ditty. It lies in the nature of a parrot to repeat all words, sentences, monologues spoken to him, inasmuch as he perceives another parrot—i.e. his own reflection in the mirror. The phonograph should speak always the same phrase &c every time it is actuated indirectly by the said contacts. Thus the parrot hears always the same words and, of course, repeats them himself soon. The lever should not be very sensitive otherwise the phonograph speaks every time when the parrot jumps, flaps his wings &c. At a certain distance below a separate contact is fixed. The bell rings if the time has arrived when the motor should be wound up.

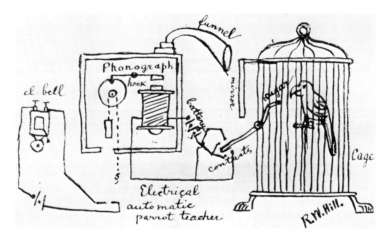

Hill's electrical automatic parrot teacher, 1894

Although Edison is credited with having thought of using his phonograph for educational purposes, such as by recording lectures by gifted teachers or distinguished scientists and so on, he probably never imagined it could be used in this way.

Not everyone had either the interest in mechanical gadgets, or the necessary skill to make them. Probably by far the most popular pastimes required no more than pencil and paper and a great deal of imagination. An absorbing geometrical puzzle dating from 1896 offered hours of challenging pleasure. It only required a number of special shapes to be cut out of card or thin metal sheet. These had then to be fitted together to make certain specified shapes—three equal squares, two unequal squares, one large square, one rectangular parallelogram, and one rhomboid—each time using *all* the pieces. The shapes to be cut were as illustrated:

Geometrical puzzle, 1896

The inventor of the puzzle, Richard Inwards of 20 Bartholomew Villas, London NW, offered a prize in the form of a copy of 'The Temple of the Andes' to the first person under seventeen who sent him a neatly drawn set of solutions. In due course someone from Aintree, Liverpool, sent in a set of solutions and duly obtained the reward. A second prize was awarded to someone from Glasgow. Do these prizes still exist in someone's attic?

This is a puzzle well worth trying and so the solutions are only given at the end of the chapter. It would probably be helpful to trace the shapes onto coloured card. In that way they can be kept the same way up. The solution is much more difficult if the pieces don't have the same face upwards.

There are definitely no prizes for a solution to the next puzzle. This one appeared in *Cassell's Family Magazine* in 1887.

The sketch is supposed to represent the ground plan of a prison having 36 cells. A prisoner in cell A is told that he can go free if he can get to B through the open doorways. The only stipulation is that he must enter *every* cell, and *must not* enter any cell twice. Apart from that he can take any route he likes and go through the cells in any order he chooses.

Three-handed draughts

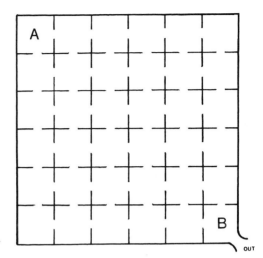

Prison puzzle, 1887

A really mind-bending game that can be played again and again is three-handed draughts (chequers). This particular game was first reported in the *Chicago Chronicle* in 1898 as the invention of George F Gruver of Covington. It has tremendous potential today as a computer game. It offers the opportunity for two people to play each other *and* the computer simultaneously. All that is required is someone to write the program. The description is as follows:

The shape of the board is substantially that of a triangle with the corners removed. It contains three sections *a*, one for each player, and a centre section *b*, connecting all three sections, and by moving over which one player can enter either one of the two sections of his opponents. The sections are sub-divided like an ordinary draught board, the alternate squares being coloured differently.

A suitable number of draughts are used—as, for instance, eleven—which are placed on the darker fields (squares) of the two outer rows, the outermost one of which is the base line.

The normal rules of draughts apply, the aim of a player being to advance with as little loss as possible over the centre field into either one of the two fields of his opponents, until he reaches the base line of a section when he obtains what is termed a 'King' which is formed by two superposed draughts, and the advantage of which is a greater scope as to moves and the power to take away the adversary's men.

The difficult feature of the game consists of moving through the centre section which can only be entered over a field *e* (the entrance). From here, to gain one of the entrances to the opponents' sections he may move to the right or left over one of the circular fields *f* or to and over the centre field *g*. In place of moving in this manner, the advance may be accomplished by jumps—that is, by jumping over an opposing man in front to an empty field behind, the jumped man being taken. Thus for instance a man may enter the centre section by a jump onto the entrance field, or by a jump from either one of the squares in the last row, and over the entrance field e on to one of the fields f provided such latter may be empty and a man occupies e, in which case this latter is confiscated by the one jumping it.

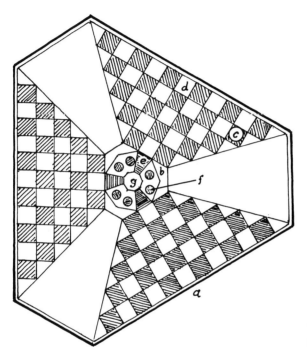

Three-handed draught board, 1898

For the central section a man may move or jump all over from field to field, and also out of it, but in this latter case only in the direction of one of the adversary's sections, and never back to the section whence it came, unless it be a King. Thus a jump out may be made from one of the fields e but not from one of the fields f or vice versa, nor shall the centre field g be jumped from one of the fields f. These rules are subject to modifications as may be made by the players. The board may also be used by only two players, in which case the unused section is simply barred to either one.

If that leaves the reader totally baffled, it is not really surprising. However, the normal rules of the game apply, the only tricky part being the negotiation of the centre section. Probably Mr Gruver's explanation makes the game seem far more complicated than it actually is to play.

'Think of a number'

An arithmetical puzzle which really does work, called 'Think of a Number' was published in the *Encyclopaedia Britannica* in 1810. The trick requires six strips of card with numbers marked on them as shown in the illustration, each strip being identified with a letter. The explanation is exactly as printed, the ƒ for s making it slightly difficult but rather amusing to read—especially aloud.

A	B	C	D	E	F
1	2	4	8	16	32
3	3	5	9	17	33
5	6	6	10	18	34
7	7	7	11	19	35
9	10	12	12	20	36
11	11	13	13	21	37
13	14	14	14	22	38
15	15	15	15	23	39
17	18	20	24	24	40
19	19	21	25	25	41
21	22	22	26	26	42
23	23	23	27	27	43
25	26	28	28	28	44
27	27	29	29	29	45
29	30	30	30	30	46
31	31	31	31	31	47
33	34	36	40	48	48
35	35	37	41	49	49
37	38	38	42	50	50
39	39	39	43	51	51
41	42	44	44	52	52
43	43	45	45	53	53
45	46	46	46	54	54
47	47	47	47	55	55
49	50	52	56	56	56
51	51	53	57	57	57
53	54	54	58	58	58
55	55	55	59	59	59
57	58	60	60	60	60
59	59	61	61	61	61
61	62	62	62	62	62
63	63	63	63	63	63

The fix flips being thus prepared, a perfon is to think of any one of the numbers which they contain, and to give to the expounder of the queftion thofe flips which contain the number thought of. To difcover this number, the expounder has nothing to do but to add together the numbers at the top of the columns put into his hand. Their fum will exprefs the number thought of.

Example. Thus, fuppofe we think of the number 14. We find that the number is in three of the flips viz. thofe marked B, C and D which are therefore given to the expounder, who on adding together 2, 4 and 8 obtains 14, the number thought of.

This trick may be varied in the following manner. Inftead of giving to the expounder the flips containing the number thought of, thefe may be kept back, and thofe in which the number does not occur be given. In this cafe the expounder muft add together as before the numbers at the top of the columns, and fubtract their fum from 63; the remainder will be the number thought of.

The flips containing the columns of numbers are usually marked with letters on the back, and not above the columns as we have expreffed them. This renders the deception more complete, as the expounder of the queftion knowing beforehand the number at the top of each column, has only to examine the letters at the back of the flips given him, when he performs the problem without looking at the numbers, and thus renders the trick even more extraordinary.

In case the way the trick works is not immediately obvious, it relies on a clever combination of arithmetic and geometric progressions. The numbers are grouped in sets according to the numbers at the head of each column (i.e. singly in column A, in twos in B, fours in C and so on). After each set there is a blank equivalent to the number of figures in each set. For example in column E, each set has 16 numbers, the first set ending at 31. The following set begins at 48, leaving a gap of sixteen numbers between the two sets. The numbers within each set are in arithmetic progression, while the numbers across the tops of the columns are in geometric progression.

Those who really like to experiment with numbers might like the following problem submitted to the *English Mechanic* in 1894:

What is the smallest number which, divided by 2 will give a remainder of 1; divided by 3, a remainder of 2; divided by 4, a remainder of 3; divided by 5 a remainder of 4; divided by 6, a remainder of 5; divided by 7, a remainder of 6; divided by 8, a remainder of 7; divided by 9, a remainder of 8, and divided by 10, a remainder of 9?

Another manipulative arithmetic puzzle was the invention of a William Radcliffe of Andreas School, Isle of Man, in 1895. He claimed to have first sent it for comments to the secretary of the Society of Arts who advised him to send this 'ingenious mathematical puzzle' to the *English Mechanic*, which of course he did. The puzzle is in the form of a hexagon containing 19 boxes or cells. The idea is to arrange whole numbers, from 1 to 19 inclusive, in these boxes so that when they are added up along specified directions in turn the total always comes to 38. These specified directions are shown by dotted lines in the diagram.

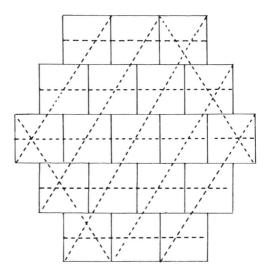

Radcliffe's '38' puzzle or magic hexagon, 1895

Mr Radcliffe obviously thought his puzzle had commercial possibilities, so much so that he prepared it in a form suitable for sale and advertised it in the 'Sixpenny Sale Column'. Seeing that he had already offered it to the world for the price of the magazine (twopence), or less if you read someone else's for nothing, he was being rather optimistic to think he might make any money out of it. The solution is given at the end of the chapter.

A puzzle sent in to the *English Mechanic* in 1894 by 'A Fellow of the Royal Astronomical Society' is an interesting and somewhat unusual conundrum in the form of a little story, as follows:

There came three Dutchmen with their wives to see me. The men's names were Hendrik, Claas, and Cornelius. The women's Geertrüng, Anna and Katriin. I forget, however, which of these was each man's wife. They had been to market to buy hogs. Each person bought as many hogs as he (or she) gave shillings for one hog. Hendrik bought 23 hogs more than Katriin, and Claas bought 11 more than Geertrüng; likewise each man laid out three guineas [63 shillings] more than his wife. Which was each man's wife?

A great deal of head scratching on the part of the readers of the *English Mechanic* was caused by a problem sent in by a farmer in

1900, although it probably originated from very early times. The problem is to determine what length of tether will allow a horse to graze exactly half the area of a circular field when one end of the tether is attached to a point on a railing which goes all round the field. The diameter of the field was given as 1000 yards. Remember the tether starts at the horse's mouth and ends at a point on the circumference of the circle.

A puzzle which this writer well remembers as a boy is the 'Shunter's Puzzle'. It probably originated in the earliest days of steam locomotives. This particular version was sent to the *English Mechanic* in 1892 by a correspondent signing himself 'G.W.R.' which may or may not have signified some connection with the Great Western Railway (or God's Wonderful Railway as it was colloquially known).

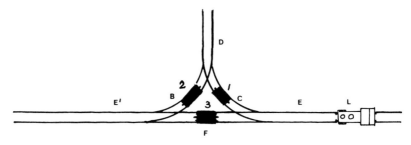

The shunter's puzzle, 1892

The procedure starts with the locomotive L facing a certain direction at E on the main line. The aim is to reverse the direction in which the locomotive is facing by shunting it between E and E', but returning it to where it started and leaving the waggons numbered 1, 2 and 3 in their original positions. No more than two waggons can be moved by the locomotive at any one time. Unlike the escaping prisoner puzzle, this one is solvable. The explanation is given at the end of the chapter.

For anyone getting bored with games and puzzles indoors, there were always all kinds of sporting activities to do outside. On a fine day, for those who could afford it, it might be tempting to go for a bicycle ride. Here is a machine devised in 1869 specially for the beginner, guaranteed not to fall over, and it even boasted power assistance up steep hills.

The machine is kept upright by a hot-air balloon, heated by an oil-fired burner at *a*. No doubt its anonymous inventor meant well but if it was ever built and tried out it must have been just as much of a menace to the rider as it was to other road users. It is a marvellously appealing idea though.

Balloon bicycle, 1869

A most exciting sport in the winter time, costing less than half a crown [12½p] was 'Sailing on Skates'. This was the nineteenth-century equivalent of wind-surfing. The following description appeared in the *English Mechanic* in 1895:

Procure a stout piece of unbleached calico, about 6 feet by 4 feet deep, fasten each end to a stout broomstick with a lacing and get a 9 foot spar to extend it. Seize a ring to the centre of each broomstick to run on the spar. Insert eyelet-holes or reef-points about 18 inches from the after broomstick for reefing purposes. To beat to windward, hold spar on right arm like a lance in rest, left arm grasping it behind the body. To go about or tack, let go right arm and the sail will fly in rear of the body to same position on left side, now grasp spar in rear of body with right hand until the next tack, when the process is reversed. To run before the wind, hold diagonally across body with both hands. To stop, let go one hand only, and hold it edgeways like a kite. Never let go both hands, or you may skate over sail and have a bad fall. When sail is on right side keep right foot in advance and vice versa. If two men are using one long sail they must turn at the same time . . . it is a most exciting sport in a stiff breeze with good ice.

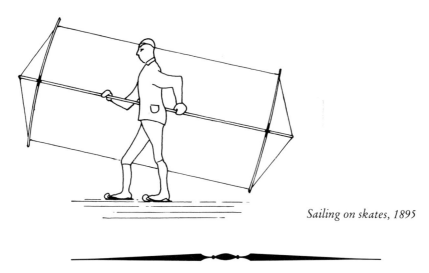

Sailing on skates, 1895

A LECTURER'S EXPERIENCE

Talking the other day with an able and popular lecturer, who seldom failed in getting a full and attentive audience, we ventured to ask him what kind of subjects and what style of treatment he thought the public liked best. 'Well', was his reply, 'I hardly know; but there's one secret I've found out—what the public hates is information.'

English Mechanic, 1866

ANSWERS TO PUZZLES

● Geometrical puzzle (p. 124)

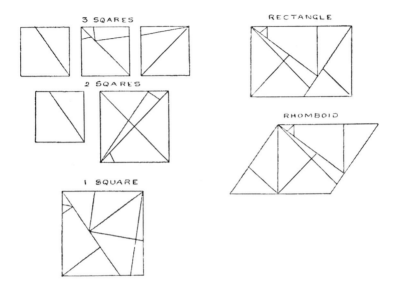

● Smallest number divided by 2 etc (p. 128) is 2519

● Radcliffe's '38' puzzle (p. 129)

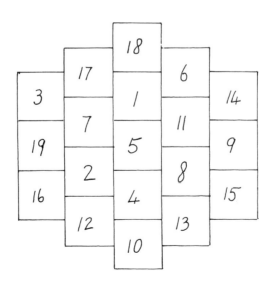

- Three Dutchmen and their wives (p. 129)
  Geertrüng buys 1 hog and spends 1 shilling
  Cornelius buys 8 hogs and spends 64 shillings
  Anna buys 31 hogs and spends 961 shillings
  Hendrik buys 32 hogs and spends 1024 shillings
  Katriin buys 9 hogs and spends 81 shillings
  Claas buys 12 hogs and spends 144 shillings.

  Geertrüng is the wife of Cornelius
  Anna is the wife of Hendrik
  Katriin is the wife of Claas

  The price paid by each purchaser must be a number of shillings equal to the *square* of the number of hogs he or she purchased. Each husband pays 63 per cent more than his wife so 63 must be the difference of two squares and must be the product of the sum and difference of the two numbers of hogs.

- Tether problem (p. 130)
  This is an extremely complicated problem since it involves the area of intersection of two unequal circles, one of radius 500 yards (the field) and one of radius t, the length of the tether, the value of the area being half that of the field. The answer is $t = 579.4$ yards, approximately.

- Shunter's puzzle (p. 130)
  Run the locomotive up to waggon No 1 and pull it back onto the main line.
  Push waggon No 1 up to No 3, leaving branch C clear.
  Run the locomotive up to D and back down branch B to waggon No 2.
  Push waggon No 2 onto the main line at E' and leave it there.
  Push waggons 3 and 1 along the main line to E, clearing branch C.
  Run locomotive up branch C and down B to main line.
  Pull waggons 3 and 1 back between the branches of the Y.
  Hitch onto waggon No 2 and pull it back to its original position.
  Run the locomotive up to D and back down C to the main line and it will now be reversed compared with its original direction.
  Pull waggon No 1 out and push it back up branch C.
  Return the locomotive to E from where it can set off facing the opposite way to that when the procedure started.

Chapter 4
Medical Science

Take care of your health. Imagine Hercules as an oarsman in a rotten boat: What can he do but by the very force of every stroke expedite the ruin of his craft? Take care of the timbers of your life-boat.

Professor John Tyndall, FRS
(addressing students at the University of London)

FOR some strange reason, people during the nineteenth century were terrified of being buried alive when they had supposedly died. Why this should have been is not known to this writer, but it does at least suggest that the population at large had little faith in the ability of medical practitioners to tell with absolute certainty when anyone was dead. At one time it was claimed that as many as 1 in 200 ended up buried alive. The rather dubious evidence offered was that it was 'a common thing to find bodies and skeletons in old churchyards that are turned over—that is to say, lying on the face'. Various proposals to avoid such an awful catastrophe were put forward, on so called humane grounds, including for example filling the coffin with sawdust or bran. Cremation too was put forward as an absolutely certain method of avoiding 're-animation'. A correspondent to the *English Mechanic* in 1897 by the name of J T Bowden cited a particular example of wrongly diagnosed death and offered a rather gruesome method of making sure that corpses were thoroughly dead.

LIFE AND DEATH. A REMEDY

While this important subject is again before the readers of the 'E.M.' it may not be out of place to give the answer a medical friend gave me to my question, some years ago, 'Have you ever met with a case of a person said to be dead who was not?' The following was his reply:—

'Yes; old Mrs —— of this town. She had recurrent illnesses of two or three times a year, when I used to be sent for to attend her case—nothing very serious but somewhat troublesome. Well, I had been attending her as usual, and had paid a visit on Saturday evening, stating that I would call again next morning. I went, rung the bell, and the door was opened by the old lady's daughter. I entered, placed my hat and stick on the hall table, when the daughter said 'Mother is dead'. 'Dead!' I exclaimed, 'When did she die?' 'Two o'clock this morning,' she said. 'Dear me! I saw nothing last night to indicate her end. I must see her so as to certify'. I proceeded upstairs, followed by the daughter. There, sure enough lay the old lady, washed and put into a clean night-dress; the blinds were all down (which I had not noticed on reaching the house) and the bedroom had been generally tidied up, and what was more she had been properly 'laid out' and quite ready to be placed in her coffin. Somehow I could not make up my mind she was dead; I listened, looked into her eyes, and did what I could to satisfy myself. I removed the head-bandages, and, after a few minutes I said to her daughter, 'I don't think she is dead.' 'Well', said my friend, 'to cut the story short, the old lady lived for three years after that!'

Such cases are appalling; but happily the above was discovered in time to save the old lady from being buried alive. With such facts before us the House of Commons should be called upon to pass a resolution that immediately before the coffin is screwed down the doctor shall be required by law to re-examine the supposed corpse, and where no certain signs of death have set in, after the usual reasonable time has expired for keeping the dead, to pass a stiletto into the heart in the interests of humanity, and in no case should such piercing of the heart be deemed murder when done in the presence of an inspector of police and certified before a magistrate of the district.

Apart from their morbid preoccupation with death and with the desire to leave this world with an ostentatious funeral, the Victorians seem to have had two other concerns. One was the state of their teeth and the other was inner cleanliness. In the days of uncertain anaesthetics and primitive instruments, a visit to the dentist must have been a frightening prospect and it was only natural to look for ways of avoiding such an ordeal. For those who couldn't afford to go to the dentist anyway, effective alternatives had to be found to get rid of toothache. Some homely rules for the care of teeth appeared in *Mothers and Daughters* in 1890. They still hold good today:

The most efficacious method of getting the mouth clean with a brush is to follow up the horizontal brushing with a perpendicular brushing from the gum downwards between the teeth. A Hindoo friend always keeps a piece of liquorice stick in his mouth while he is performing his toilet, and this piece of fibrous stick, which he keeps on chewing, acts as a splendid tooth brush. The chewing of sugar cane by the West Indian negroes helps to account for their beautifully clean teeth, and, generally speaking the world over, every animal keeps its teeth clean by chewing some sort of fibrous material. The instinct is born in children who love to chew bits of straw, grass and stick, and are often reproved by parents who little know that this is one of Nature's lessons to her little ones.

The modern habit of using all foods sloppy or artificially prepared has much to answer for in the way of dirty teeth. Notice the difference between such a simple thing as whole-meal bread and white bread! The former is much less likely to stick between the teeth, and the flakes in it have a scrubbing action on the enamel, which white flour bread has not. The black bread of the peasants of other lands, and of England in bygone times, not only from its chemical constituents, but from its mechanical action, did much for the preserving of the teeth, and as a result the best teeth are not to be found among the higher classes who take the most artificial care of them, but among those peasant races that live on the hardiest teeth-cleaning foods. Quite an unsuspected cause of dental decay is the use of flesh foods and soft starch foods. The fibres of the flesh get between the teeth, and there rapidly decay. This constitutes the great difference between the fibres of meat and the fibres of the liquorice root. The latter cleanse and do not decay, the former decay and do not cleanse. The best thing to do is to see that the daily food contains something or other which will give teeth work of a cleansing character. A thick piece of wholemeal bread is fairly good; but the chewing of liquorice root, or sugar cane, or some other fibrous substance (like tough celery) is still better. If using a toothpick, use a quill or a bamboo splint, or a thorn from a hawthorn bush. Don't use pins or needles, or metal or any sort.

Some sectors of the community of course didn't have much choice in what they had to eat. They were the ones for whom a visit to the dentist would have been unthinkable. They were expected to fend for themselves—and usually did. A way of making one's own fillings in 1879 was as follows:

Take an old silver thimble, or other silver article and with a very fine file convert it into filings. Sift through gauze—to separate the coarse from the fine particles. Take the finer portion and mix with sufficient quicksilver to form a stiff amalgam, and while in this state fill the cavities of the decayed teeth. As it turns black under the action of the acids of the mouth, it should be used sparingly for front teeth. A tooth should never be filled while it is aching.

Then, as now, the best material for fillings was gold. The way to make gold fillings was explained by a 'Dental Mechanic' (1879):

A good metallic stopping is made of the following proportions:— 12 grains of pure gold; 10 dwt of pure silver; half the weight of the two together of pure tin. Place the whole in a crucible and fuse to a liquid, then procure some fine sand, moisten and place in a small box or pot; pierce a hole in the sand large enough to hold the metal; pour it in and when cold, granulate by filing with a coarse rasp. When wanted for use place a little in the palm of the hand, according to the size of the cavity, then add sufficient mercury to make it adhesive; place it in a piece of linen and press all the excess of mercury away; it is then ready to place in the tooth.

What the dental mechanic neglected to say was how to actually get the filling into the tooth. The necessary instructions appeared in the *English Mechanic* 21 years later, in 1900. Presumably the original querist was past caring. What is more it hardly seems possible for anyone to do the job without help. However, this is what one was supposed to do:

Get a piece of straight-grained bundle firewood and split into sticks about $\frac{1}{8}$ inch square. Reduce the ends that you can get it into hollow of tooth. Now take a small portion of clean cotton-wool and clean the tooth well out and dry it. Now take another piece of wood, and make it so that you can get it into the tooth with the end square. Now take a small portion of nitrate of silver (about as large as a dressing pin's head); wet the end of the fresh stick and pick up the nitrate and insert it into the tooth and ram well home; then take a small quantity of cotton-wool and saturate it with a strong tincture of myrrh and plug it up.

Don't swallow your saliva for a time; if done properly that tooth will never trouble you again. See that you don't drop the nitrate of silver down your gullet.

As for ensuring 'inner cleanliness' a form of self-inflicted torture was to be found in the treatment of constipation. Hydropathic methods were particularly 'popular' as also was electricity. This example, dating from the 1890s, came (unusually) from a woman correspondent to the *English Mechanic*.

You need have no fear of trying the full gallon of water. I have used the treatment for over five years. At first, morning and evening, in a hot bath, 110°, using the same water to fill the enema cistern which was hung on a nail on the bathroom wall, as high up as the 8 ft rubber tube would comfortably allow. Where there is not a bath, place the cistern on a shelf, or hang it where you have *height to give power*. A hand enema may be used but is not so comfortable. When you once get the colon not only emptied but cleansed, there is still the rest of the numerous smaller channels to be purified. This takes time and patience . . . .

Another correspondent drew attention to cocoa as a cause of constipation but warned that using enemas—especially with *cold* water—led to haemorrhoids. In that event, a sure fire cure was offered by someone calling himself 'The Kent Hermit' which may or may not be significant. This was in response to a query from 'Drysdale' in 1900.

There is a perfect cure for bleeding piles; but its action cannot be explained, and it is so simple that it may not receive the attention that it merits— especially from people who do not fancy being cured for nothing. The remedy is to wear a horse chestnut on each groin. If 'Drysdale' be a man, let him carry a horse chestnut in each front pocket of his trousers, as they cover the groins. In a little time his system will absorb the curing essence of the nuts, and the disease will gradually become cured. Avoid all food and habits which aggravate and foster the disease. Be careful to use only horse chestnuts because Spanish chestnuts are of no avail.

This raises all sorts of speculation about the validity of the Kent Hermit's experimental method. What did he do to reach such a firm conclusion? Was it the Spanish chestnuts that ruined the results and forced him into becoming a hermit for 'social' reasons? One way of avoiding the problem altogether would of course be not to go for cold water enemas and not to drink cocoa.

A very effective alternative form of medication is electricity. Medical uses of electricity in the nineteenth century were legion and more than likely lethal. The following account by a man named Gerard Smith appeared in the *English Mechanic* in 1895:

In the practical use of electricity for disease, it is, after all, upon emipirical lines we have to proceed. Theories about the direction of current, the chemical or mechanical or vital action, the direct or reflex nature of the effects, and many more, are pretty and attractive; but as yet we have attained our best results from simply trying, almost haphazard, various applications in various cases, and noting which seems to do most good.

There are some general applications of electricity—I mean applications to the whole body—which have indubitably good effects, general faradisation being the most valuable—that is, the application of the secondary interrupted current from a coil to the surface of the body, best attained by the electric bath. The refreshing and stimulating effects cannot be doubted, though possibly a similar vibratory stimulation of the skin surface by other means would also result in good. Faradism (as the medical application of the secondary is called) is certainly the most convenient method of producing a thorough rousing 'prickling' over the surface of the body, and thereby stimulating a lazy or sluggish nervous system to more lively action.

The bath may be taken by placing the poles [the electrodes] at either end of the bath, insulated from metallic contact with it, and from the patient, whose body simply takes the place of being part of the conductor, or one pole may thus rest in the water, and the other be passed over the body by means of a metal plate and handle.

This is a stimulant action and must be avoided in cases where sedation is required, such as for pain or spasm. Another use for faradism is for lazy intestines. Here it is best applied by one pole, a smooth metal rod placed in the rectum, and the other by means of a large metal plate wrapped in a wet towel firmly pressed on the abdomen; a strong current should be applied, used daily for ten minutes. This beats all the pills and potions in many cases, especially if followed by kneading deeply the intestines, and a full enema of hot water.

An excellent method for the general waking up of nervous and depressed folk, who think too much of their symptoms, and need a good shake up, is the electric douche. The patient stands naked on a metal plate attached to one pole of a strong coil, and a small but powerful stream of cold water from a pipe attached to the other pole is made to play freely over the body. These patients, who handle their bodies so tenderly and with such great respect, are often greatly roused and benefited by such treatment applied with no respect at all.

Mr Smith goes on at great length discussing all the varieties of applications of electricity, including loss of voice, cured 'by means of a fine wire brush to the throat ...', sexual problems, rheumatism, and gout. He states that Faradism also gives 'artificial gymnastics' to paralysed limbs and it is useful in removing blood clots in the brain. 'Very weak currents only (5 milliamperes or so) must be used to pass through the brain .... The spinal cord will bear greatly stronger currents, up to 30 milliamperes.' He also claims that electricity can be used to force substances through the skin into various internal organs where medication might otherwise upset the stomach if it had to be swallowed.

From the casual way that electricity was treated, it is clear that people just did not appreciate the dangers. This was especially so in domestic installations where frequently wires were nailed to walls and even attached to gas pipes which provided a convenient 'run'. *The Lancet* in 1896 published a report of an accident with electricity in which an electrical engineer gave himself a 2500 volt shock yet astonishingly survived, and declared himself to be 'decidedly better in general health'.

On Nov. 20th [1896] an electrical engineer was standing on a chair ready to move a reversing switch on the arc lamp side of a rectifier for lighting a series of sixty lamps (3000 volts). He casually placed his hand on the metal cover of an ammeter (which was in series with another sixty lamp circuit) and turning to the left and leaning back to make an observation (still standing on the chair) he placed his left hand on an iron pillar. Now the ammeter case proved to be in metallic contact (accidental) with the light leads, and as the return wire was 'earthed' there played through the body from hand to hand a current under a pressure of (nominally) 3000 volts. As a matter of fact, the electromotive force tested between the points of contact by a Kelvin electrostatic voltmeter proved to be 2500 volts.

He says that the first thing he realised was that he found himself standing on the floor, but has no clear idea as to whether he jumped off or was 'knocked off'. The forearm was drawn up close to the chest and the hands clenched. From a little above the elbow downwards he described a 'feeling of pulsation' and a violent beating in exact step with the alternators (running eighty-three periods per second). All power of movement below the elbow was absolutely lost, but the arm at the shoulder could be moved from the side. The pulsations, although still retaining their periodicity, soon became less violent, and motor power returned successively in the muscles that move the elbow, wrist and fingers. In three minutes he felt 'none the worse'. There was no sensation of burning at the moment, but ten minutes afterwards the hands became painful, and examination showed a burn on the tip of the middle and ring fingers (L) and on the back of the little and ring fingers (R) with a seared line across the lower portion of the palm. There has been no other effect excepting that he expresses himself as feeling 'decidedly better in general health'.

No doubt his feeling about better health was more to do with finding he was still alive, than for any recommendation to others to try electric shocks as a stimulant. One interesting question arising out of this account is how did he count the frequency of pulsations in his arm and how did he know they were in exact step with the alternating current frequency? Related to that it is interesting to note the rather curious frequency of 83 cycles per second, compared with the modern rate of 50 (Britain) or 60 (USA). A correspondent to the *English Mechanic* of a rather more cautious disposition than the unfortunate electrician, advocated wearing two pairs of cork socks and boots made sufficiently large to accommodate them as a means of avoiding electric shock.

Anyone suffering from constipation and other 'disorders of the intestines' who didn't fancy either the water treatment or the electrical alternative, could still turn to straightforward medication—especially if he or she happened to have some creosote available, or some coal to spare. The use of creosote in medicine has a very long ancestry—at least 400 years. The following account appeared in 1898 in a journal called *Cronica Medica*:

Vladimiro de Holstin finds in creosote an excellent means of combating chronic constipation without exercising any purgative action properly so called. The creosote should not be prescribed in pills, capsules, or alcoholic solutions, but pure and in drops. The effective dose is about seven or eight drops, taken twice daily, immediately after breakfast, and after dinner, in a glass of milk, beer, wine-and-water, or pure water. To begin with, one drop of creosote is administered, and that amount increased by one drop daily until the desired effect is obtained. In this way the necessary dose is determined for each case individually. This treatment has to be continued for some time—some months in fact—and not only overcomes the chronic constipation, but at the same time restores the appetite and braces up the system.

In a similar vein, but more than twenty years earlier (1866) an article appeared in *Analysis, Berthier* describing experiments in administering anthracite by the medical officer to a regiment of Hussars in Germany.

Dr N Dyes, chief medical officer of the Hanoverian regiment of the Hussars of the Guards at Verden, having remarked that pigs eat coal with much avidity, took the idea of adding a certain quantity of this mineral to the food given to fatten them. He gave to some of them from three to six drachms per day, and remarked that these were noted for their liveliness, good appetite, growth, and rapid increase of flesh. Having repeated the experiment on several occasions, Dr Dyes came to the conclusion that in the fattening of pigs anthracite is, for the greater number of these animals, superior to common salt as a means of quickening digestion and maintaining general health; that it constitutes a preservative against catarrh of the stomach and intestines, as well as against cholics, and that it prevented all the maladies which result from the impoverishment or thinness of the blood; also that coal exerted a favourable action on the liver, spleen &c. Perfectly convinced as to the correctness of his observations, Dr Dyes did not hesitate in administering coal to persons attacked with abdominal complaints, and chose for that purpose the Piesberge anthracite, which is found in great quantities in the neighbourhood of Osnabruck. He obtained invariably and rapidly the best results from the employment of this mineral, which had not in any case produced the slightest evil effect. For two years Dr Dyes has employed anthracite in a number of abdominal affections, and obtained such remarkable results that he declares that it cannot be too much recommended in catarrhs of the stomach or cramps, and a host of other maladies which are usually treated by coal tar or charcoal. For diseases of the skin he prefers anthracite internally to coal tar externally. Dr Dyes gives in fine powder 15 to 30 grains per day, incorporated in some convenient pulp, or in pills. There is nothing extraordinary in all this when we consider that anthracite contains carbon, 0.90; hydrogen, 0.03; volatile matters (sulphur) 0.03; ash, 0.04 in 100 parts.

A comment on 'unusual' substances used in medicine appeared in the *English Mechanic* in 1897 under the heading 'Explosives We Swallow'.

Prof Alonel says that we often swallow or apply substances which, if incautiously treated or used in any but the minutest quantities, would blow us to atoms. What is more, these substances, so destructive in large quantities are of the most beneficial nature when used in the form of medicine. One of the best remedies for heart trouble, neuralgia, asthma, and headache is nitro-glycerine, which is the only explosive ingredient in dynamite. The dose is only one two-hundredth of a grain dissolved in spirits of wine, or combined in gelatine tablets. Collodion, a syrupy-looking liquid that is used to form a false skin over abrasions of the cuticle is nothing but gun-cotton dissolved in alcohol. In its natural form it is one of the most dangerous of explosives, and yet, as a medicine it has no equal for the purpose which it is used. Another explosive used as a drug is picric acid. This is prepared from carbolic acid and is administered internally in very small doses for ague and headache. This acid is one of the explosives used in the preparation of bombs. These and many other dangerous drugs are perfectly safe when used as ordered by physicians.

As an alternative treatment for persistent headache a surgical method was described by an Irish correspondent to the *English Mechanic* in 1898.

Trephining for Jacksonian epilepsy, persistent headache, and obscure cerebral symptoms is, if anything, but too common. It will take a long time yet before that wondrous organ the brain is thoroughly understood; there are cases on record where large foreign bodies such as pieces of iron 2 inches or 3 inches long have been imbedded in it as a result of an accident, without affecting the individual mentally or physically; while on the other hand, a slight depression, a minute spicule of bone, a small bone tumour on inside of skull, or a small localised thickening of the delicate membranes lining the skull, may convert a useful member of society into a staring imbecile. Several mental and physical symptoms may follow small localised continuous pressure on a part of the brain, producing chronic irritation, showing itself as a sensory product in continuous headache, or as a motor product in convulsions, epilepsy &c. due to the compression of the brain cells by the products of inflammation, or to the interference in their working by the direct inflammation itself. The operation of trephining removes the direct cause of the disturbance, and many a case of epilepsy, compression &c. has been saved thus. I have myself seen a whole slice of bone removed from the side of a skull of a child suffering from hydrocephalus, or 'water on the brain', to relieve headache and convulsions, with much apparent—temporary at any rate—improvement, mentally and physically.

There were of course cheaper and less drastic ways of relieving a headache, especially if it was accompanied by a high temperature as a case reported in *The Lancet* in 1897 shows.

On the evening of October 19th I was called to visit a Roumanian boy, six years old, suffering from typhoid fever. I found him *in extremis*, almost pulseless. The child's head was completely wrapped over with a large white sheet, and as I looked at it, this enormous white envelope seemed to be on the move. Whilst I was surveying this covering there crept from under it a small frog, which quietly sat over the child's left eye. It seemed quite content. I immediately called the mother's attention to it, and requested her to take the beast away, thinking that it had crept there as an intruder. 'Oh no!' said the old lady, 'a doctor recommended that a lot of them should be kept to the head to keep it cool'. Seeing the head covering still on the move I raised it for curiosity, and in a second out jumped about twenty other frogs, and away they hopped in all directions. I have often heard the expression 'as cold as a frog', but this was the first time I had seen a frog applied as a head cooler.

An article in *Ice and Cold Storage* in 1896 advocated being put in a refrigerator as a means of curing all kinds of ailments—including insomnia:

The temperature of one hundred and ninety degrees of frost is one appertaining to no habitable region of the globe. This is not encountered by the stalwart and strong, who brave the regions of eternal ice and unmelting snow, but is the temperature meted out to the sick and ailing, and those afflicted with bodily disease, by a famous foreign physician—not for the purpose of killing them outright—but for the purpose of relieving morbid and unnatural symptoms, and restoring them once more to splendid and vigorous health.

The patient ... is first wrapped up in woollen garments, and is covered from head to foot in a thick fur coat. Then he is placed in a well or box, and made to stand on a wooden stool placed on the floor of it, while out of a hole in the lid only his head and fur-protected neck emerge. The box is made with a double wall, and in the cavity a freezing mixture formed of carbonic acid and suphurous acid is poured. This generates the very great cold which has been mentioned, and affects the air in the interior of the box, and therefore the patient who is standing in it.

The radiations of heat—or cold, to speak popularly—pass through the fur and through the skin, just as X-rays pass through certain metals, cloths and other opaque bodies, as well as the skin and muscles of man, and enter the body of the patient; but, as has been said, he does not feel any cold because they are stopped by the refrigeration of the skin.

As soon as the body is cooled at all it makes an effort to reassert itself and to regain its temperature, to which fact is due the after-glow which follows the morning cold bath. Under the influence of this fearful cold the body makes a tremendous effort to recover itself from the shock of which it has had no ordinary warning. The result is that almost immediately the blood begins to circulate with increased energy; the pulse gets quicker; the body tingles and quivers with a remarkable glow, and the physician is able, by means of a thermometer, to see that the temperature of the patient actually begins to rise. After ten minutes sitting in the freezing box the patient is taken out, his furs and wollen garments removed, and his ordinary clothes are given to him, and the treatment is over for the day.

Almost immediately a feeling of ravenous hunger is produced due to the using up of whatever stores of material were in the body, in order that the internal furnace, which burns evenly in every one of us, may keep up the work of maintaining the temperature, and at the same time as the sensation of hunger is produced a feeling of mental exhilaration and vigour, as well as of great physical energy.

Dangerous substances and morbid products of digestion which may be circulating in the blood are thus burnt out rapidly, and the vital fluid is cleansed thoroughly, for all the world as if it had been given a bath, and even more thoroughly than the skin is cleansed by the ordinary benign influence of much soap and water.

Dyspepsia, that bane of present-day life is in consequence one of the first diseases to succumb to this treatment, and a good many other complaints, including liver and kidney diseases of various kinds are all cured by it, as are nervous exhaustion, with its attendant complaint of insomnia. The remarkable effects of the cold on the whole nervous system are indeed very great, and melancholy and depression ... are quickly banished. Lowness of spirits is a potent cause for increasing the effects of these diseases, because the patient takes so gloomy a view of his condition, and the cold, by removing this cause as well as by exciting the whole body to a vigorous action, removes one difficulty from the treatment, and thus clears the way for a perfect cure.

Extreme cold was also advocated as a means of improving the appetite in 1899, as the following report shows:

The benefit of pure air as an appetiser, especially if taken in connection with outdoor pursuits, is well known to every Nature lover. Now it appears that air in liquid form is of benefit to the appetite from the results of lowering the temperature.

The Bulletin of Pharmacy is responsible for the statement that a Russian physician recently placed a dog in a room with the temperature lowered to 100 °F below zero by the use of liquid air. 'After ten hours the dog was taken out alive, and with enormous appetite'. The Bulletin does not state how much more 'enormous' was the appetite after a ten hours' fast in the very cold room than it would have been at normal temperature.

However, the physician was so pleased with the manner in which that dog took to his food that he tried the test himself and reports this result: 'After ten hours' confinement in an atmosphere of still dry cold, his system was intensely stimulated. So much combustion had been required to keep the body warm that an intense appetite was created. The process was continued on the man and the dog, and both grew speedily fat and vigorous. It was like a visit to a bracing Northern climate.'

If this theory is to be followed out, a reduction in the price of coal and house furnaces is to be expected. The reverse effect is true of cold-blooded animals. Prof. Smith of Yale College is authority for the statement that frogs may probably be kept alive for several years, possibly even as many as ten, without food in water of uniformly very low temperature.

Quite what this last statement means is not clear. Presumably the frogs would not actually be frozen in blocks of ice. If the water is to remain as liquid then of course it cannot be reduced in temperature below the freezing point. It would be highly doubtful whether frogs could be kept alive for ten years.

Returning to the question of headaches, brain fever and the like, in 1892 a report by an anonymous correspondent in the *English Mechanic* advocated a novel use for that universal panacea for all childhood ailments—castor oil.

Some years ago, on trying castor-oil as a means of preventing the hair falling off, I found it affect my sensitive brain in such a way that I became convinced it would be the best remedy for congestion of the brain. I have since recommended it with advantage to those who have tried it. A woman who for some years had been unable to attend to her household work owing to a terrible chronic headache was soon set right. A month or two back I was told that an acquaintance had been suffering for a week from brain fever, and that the medical man pronounced the case to be next to hopeless. I persuaded the friends to instantly apply the castor-oil, with the result that there was a good night's sleep and a recovery. Ice can be reapplied when the anointing is done—i.e. over the castor-oil. The head should be oiled again when the first application seems to be drying. It would be useful in any kind of fever. The castor-oil thins and dissipates the blood. I expect that it would be a good ointment for inflammation in any part of the body; but I have tried it for the head only. It should be used also where there is merely mental excitement, as in a healthy subject it has a depressing effect. In lunatic asylums its application as here described would be invaluable.

The next item, also about the brain, does have modern research to lend it credibility. It has been shown in recent times that people can function perfectly well with only vestigial brains. The tone of the item suggests that the discovery that a patient could live almost normally for years with much of his brain destroyed by a tumour was then (in 1897) quite revolutionary.

At the request of a number of prominent physicians of Philadelphia, Dr S S Koser held a post-mortem examination of the remains of John Bly. Bly, who was twenty years of age, for a long time suffered with a tumour which grew into the very base of the brain, and occasioned his death. The growth had a visible effect on his brain and the case became a curiosity to the medical profession. The tumour was imbedded too deeply into the brain tissue to admit of an operation. It was found that the tumour was nearly as large as a billiard ball. It was so located as to demoralise the nerves of the sight centre, and as a consequence young Bly was blind for over three years. The most singular fact developed was that the entire brain had been hollowed out by the action of the tumour. The cavity was at least 5 inches in length, and was filled with pus. All that was left of the brain was a thin shell, composed of the tougher tissues where the brain matter gathers into nerves, which were less susceptible to the progress of decay. When an incision was made into the shell the whole mass collapsed.

The circumstances which made the case almost unprecedented in the annals of medical science was the manner in which the patient retained his rationality and faculties under the circumstances. He had the sense of touch, taste, hearing, and smell, had very tolerable control over his locomotor muscles, could talk, and in fact was comparatively discommoded in no other way than by the loss of vision. His retention of memory was remarkable. He was able to memorise poems up to within two weeks of his death.

Then as now, men were concerned about loss of hair and bald patches. As we saw earlier, castor oil had been tried and led to the discovery of its potency against congestion of the brain. Other efficacious lotions of the 1880s and 1890s were

... tincture of capsicum, and equal parts of tinctures of capisum and cantharides and plenty of patience. The hair will gradually grow again but it will be an affair of months rather than weeks.

And, to cure a bald spot

... Quinine sulphate 20 gr., glycerine 1 oz., cologne 2 oz., bay rum 2 oz., rose water 11 oz. Rub glycerine with the quinine and add the other ingredients in order named. The addition of fluid extract of jaborandi is said to stimulate.

In May, 1898, Dr W G Black, FRCSE commented on a disturbing observation of his that baldness amongst doctors was on the increase. He wrote:

People who have frequented public meetings within the last twenty years or so must have had their attention drawn to the increasing number of bald pates amongst the audience on the benches. Beyond that time past one may generally have observed a large proportion of grey or white haired heads in congregations or meetings; but these now seem to have disappeared, and the bald pates have gradually taken their places. It may be hoped that this may not be a sign of the bare head becoming hereditary in the youth and manhood of the age, and that wigs will again come into wear and fashion as in old times. This decay of hair so early in youth and manhood of the present day may be due to the fashion of short cropping the head, even in boys now, as there is little or no baldness noticed in women who are not cropped short in their hair, not even as girls.

The explanation of this decay, which seems to be irremediable in maturing age, is probably due to the young hairs budding out at the side of the older growth being deprived of shelter and warmth to keep up their young vitality, and they consequently wither down from exposure. The stronger hair only can therefore survive this exposure to the weather, and it even is being stunted in growth owing to its being frequently cropped short. A like process goes on in meadow and cornfields when the coarser stubble and straw and grass allow the growth amongst them of fine grass, weeds, and flowers which would perish if left alone without any shelter.

On the occasion of visiting an old society rooms in a large provincial city, I was pressed to look over a large collection of portraits and groups of past members, and could not but notice the finer heads of hair in the older ones, and the gradual appearance of short-cropped hair and the bald pates looking in increasing numbers in the newer members. One generally associates bald heads with the legal profession, where it is owing to the custom of wearing wigs necessitating the hair being cut short....

That it should be making its appearance in the medical profession seems to be a new custom or fashion, and it probably will in due time lead to the doctors being habilitated in wigs, as our Mediaeval ancestors were in going about the towns.

It may be surmised that there may be a hygienic danger in the shortness of hair or complete baldness by the exposed scalp being deprived of its natural covering and armour, shielding it against cold draughts and vitiated air charged with microbes. Hence one may hazard the suggestion of making enquiry whether such individuals are not more liable to influenza than persons possessed of a good substantial head of hair.

This last suggestion sounds like an interesting piece of research for someone. However, there are a few worrying points in Dr Black's appraisal of the situation. One wonders what his knowledge of genetics was at that time and whether that was a topic in medical degree courses. One question which didn't seem to have occurred to him (or if it did he dismissed it from his mind) was why do men still have hair following the widespread custom of the seventeenth and eighteenth centuries of shaving the head and wearing wigs? If his theory was valid, then by now there ought to be hardly any men with a head of hair. One way to find out if he is right perhaps would be to wait and see what happens to those young people who today go about with their heads shaved or with just a few tufts remaining painted green! Curiously enough, although he seems to have displayed total ignorance about genetics, he was up to date with his awareness of the existence of microbes and the connection between microbes and disease. This is rather more than could be said of some other medical men despite the researches of people such as Pasteur and so on. An instruction given to doctors in 1874 about precautions to be taken when visiting a sick-room was rather ambiguous on the question of the sources of infection.

Never go into a room where there is an infectious disease without solid food in your stomach; if the air is very offensive take a few drops of Condy's fluid on a lump of sugar; do not stand either between the fireplace or an open window and the patient, or in any direct current of air from the body across yourself.

Malaria

If the medical profession as a whole was none too sure about what caused disease, it is excusable for ordinary people to have wrong ideas. The following item offers an explanation for the origins of malaria and jungle fever derived from direct experience arising from long service abroad. The item was written in 1892.

Under the definite term of malaria come all deadly exhalations generated by the combined action of sun, heat and water. I am a living instance of the rare few who have survived the violent fevers and their resulting sequelae, due to temperate habits, active outdoor life, and a robust constitution. Gases are the recognised causes of these maladies, pernicious in their composition and their effect varying according to the recipient condition of the human health. The liver is the organ of paramount importance to be treated on entering a pestilential locality: ten grains of calomel, followed by a purgative draught will clear it of bilious obstruction; avoid heavy meals of solid meat; and as a prophylactic febrifuge, at sunrise and sunset swallow a small wineglassful of any strong spirit in which five grains of sulphate of quinine have been dissolved. Strong black coffee, made by an infusion of $\frac{1}{2}$oz. freshly roasted coffee in a quarter pint of boiling water is also a valuable antidote at meals. For want of quinine I so treated many desperate cases of fever among my native followers with much success: it combats congestion of the brain, the worst feature of all malarial fevers.

During my 39 years residence in India I was thrice attacked by the great enemy of Europeans who follow wild animals into their wildernesses. I was twice in a critical condition, the brain being the seat of danger. I got no sleep for many days and nights, when I was bled on two occasions until I fainted, and then slowly recovered. Two of my jungle fevers were produced by several days' sojourn in districts covered with dense forests and swampy grass plains. The third one attacked me and a friend living in a comfortable dry dwelling. Nearly every human being, black and white was prostrated simultaneously by this epidemic fever, haunting a subsoil of decomposed granite, and usually appearing every third year. Quinine was contra-indicated in this severe form of malady, as it increased the cerebral symptoms; but port wine, with Fowler's solution of arsenic, and the powder of Aconitum Heterophyllum were substituted with success. Ten days' excruciating headache with constant vomiting and purging were succeeded by much-enlarged spleens. As soon as able to quit our beds, we took our departure on wheels from that pestiferous spot, and by the aid of much porter, with citrate of iron and quinine, became convalescent in a fortnight.

Mosquito curtains, from sunset to sunrise, have been found effective screens from the heavy poisonous vapour by persons obliged to live in a tent.

The same correspondent was back five years later with a letter headed 'The Therapeutic Value of Venomous Secretions' again a topic which has relevance to modern research.

About 20 years ago an old friend, a beekeeper, told me he had tested the German discovery of stings applied to rheumatic swellings on his feet—some three or four of the insects were made to inject their poison on the seat of the affection—and he was cured. Recently [1897] a Government medico in Western India is reported by the last mail to have met with some success in severe cases of bubonic plague, by hypodermic injection of dilute snake venom. Lastly, I think, the acrid secretions of Indian toads and some of the lizards, may yet be found to possess peculiar properties. As yet these are only utilised by the ivory workers of the N.E.Frontier, Bengal, one of whom told me the secret of softening the long strips cut from elephants' tusks for weaving those wonderful mats spread near the thrones of barbaric chiefs. After being sawn into the proper sizes, these ivory pieces are packed in bags with live toads, until the material becomes soft and flexible. The same man also worked skilfully in buffalo horn and rhinoceros' hide—rendering the latter semi-transparent. I believe those liquids secreted by toad and lizard in India are compounds of ammonia and phosphoric acid. I knew a lady who carried a deep scar on her neck from the ejection of fluid by the latter reptile, suddenly disturbed. Some 60 years ago the Rev. Skrimshire, a Norfolk clergyman, a zealous entomologist, discovered that the yellow juice which exudes from ladybirds was a specific for toothache, and his son, a chemist at Ramsgate used to sell the extract. These same insects one year caused great mortality among horses and other pasture stock; they swarmed on the grass in myriads.

Toad venom was also advocated by a Mr A W Leslie-Lickley of Thornton Heath, Surrey, as a cure for epilepsy. He reported an amusing anecdote about a woman who tried to poison her husband with toad venom and thereby discovered a cure for dropsy.

As illustrating the power of the toad venom over dropsy, it is related that the husband of an Italian woman was dying of dropsy, but taking so long about it she thought to help him on. She accordingly procured a toad and put it in his wine so that he might drink the liquid and die; instead of doing this, to her astonishment he completely recovered.

Another potential source of infection was the common earthworm. This came under scrutiny in a report published in 1898 relating to the views of a Dr Halstead Boylan.

Dr Halstead Boylan of Paris has recently presented some ideas concerning the cause of yellow fever, charbon [anthrax] and tetanus, and he claims that the worms coming to the surface of the earth from the bodies of those who have died of these diseases frequently act as propagators of the germ or poison. The earth itself becomes contaminated and the poison is inhaled by man, eaten by cattle, or infects the water which afterwards becomes a fruitful source of disease propagation. Verneuil has called attention to the fact that tetanus is most frequent with those coming into contact with horses, it only being necessary for men under these conditions to have upon their hands a small cut, scratch or wound inadvertently brought into contact with the manure containing the bacillus or with the ground infected thereby. The first experiment of this kind was made by Nicolaier who demonstrated that the subcutaneous inoculation of earth taken in the fields, gardens and streets determined in mice, rabbits and guinea-pigs at times septicaemia by septic vibrios, and at others tetanus.

The contamination from below is twofold: miasma and emanations from the soil in paludial districts, or when particles of organic substances rise to the surface in different states of alteration, coming from the tissues of decomposing animal matter and by the ordinary rainworm.

Pasteur has found the germs of charbon in the superficial layers of the earth over ditches where sheep that had died of the disease several years previously had been buried. The earthworms are the first carriers. In the depths of the soil they swallow minute particles of earth containing the spores formed round the bodies of men and animals for the purpose of extracting from them what nutritive substances they may contain, and rendering them on the surface in the form of little twisted cylinders of a dark brown colour. The practice of cremation is strongly urged to destroy these dangerous elements of both the bodies of those having died of the contagious and infectious diseases and the soil surrounding their place of burial.

Despite Edward Jenner's pioneering work on vaccination in the eighteenth century, arguments still raged a hundred years later as to whether vaccination was of any use or not in combating infectious diseases. In the 1880s various Acts of Parliament came into force with the aim of improving the nation's standard of health through improved sanitation and the demolition of slum dwellings. At the same time experiments were carried out in the large scale vaccination of residents in various centres of population. The observed improvements in health were claimed by the proponents of vaccination to be due entirely to that, while those who thought vaccination was disgusting argued that better public health was entirely due to better drains. A Mr Alex Wheeler was one of the latter. He published his views in the *English Mechanic* in 1892, using as the basis of his argument statistics gathered from experiments in mass vaccination against smallpox in Sheffield.

Sheffield has in the last thirty or forty years been as much vaccinated as is practicable: there are limits to the full vaccination of the population. This town thus vaccinated has been visited with a considerable number of smallpox epidemics. These epidemics have been most prevalent in one—and that the oldest—part of the town. The worst have either begun, or soon settled down in this quarter. . . . The best parts of the town have been left comparatively clear of the small-pox. It loves the close, ill, or unventilated places, where the poorest dwell. The part I speak of is in the centre of the town, is about 135 acres in extent, and was condemned by the medical officer in 1877, and recommended to destruction and rebuilding. Only a very small portion has been rebuilt. The rest, up to a very recent date was almost unaltered. This is the part that figured for the heaviest sufferer in the small-pox; it was in this part the largest number of deaths occurred. It contains the huge midden system of the town, and this is the thing that has been condemned by the officials for long, long years, till now it is in the way of removal as far as possible. The deaths in this part are so heavy that if the rest of the town was the same, there would be 4000 more to bury every year. This, and not the vaccination or non-vaccination of the people is the reading of the small-pox in Sheffield. It is, as I have pointed out in Darlington and Paris, a question of conditions of living, and not of vaccination. In Sheffield the rich had not as much public vaccination as was done in this quarter, yet the rich did not yield a single death by the small-pox in the last epidemic. The bulk of the deaths were of poor folk. Of 7000 cases of small-pox, it is allowed that 5800 were vaccinated . . . the vaccinated had more deaths than the unvaccinated, and six times the cases. There was every kind of small-pox among the vaccinated, every type of it, and the only exemption in the town was of those in good circumstances; so that good circumstances was the protection, and not vaccination. I fear until conditions of living of the poor are changed small-pox will be found among us, and as now, if we continue vaccinating it will be mostly among vaccinated. In the London Hospital there are now regularly more than 94 per cent of the cases vaccinated.

I therefore submit that the rite being proved by all experience a total failure as a protection, it is as foolish as it is illogical to ask us to plead for the abolition of compulsory vaccination on the ground of its injuries alone. It is the utter failure of the fetish which makes me wish to be rid of the national enforcement of it. I care not who is so credulous as to adopt it. What I do care for is the enforcement of a useless and injurious rite. With the adoption of the Public Health Acts, coincident with a decline in the adoption of vaccination, we have had a decline in the fevers, small-pox included. The universal experience is that sanitation is the remedy. In too many homes vaccination has been found wanting, and the day of its wane is near.

One could be forgiven for wondering what all the fuss was about, and there is no doubt that Mr Wheeler was 'massaging' the statistics to support his view in just the same manner as the pro-vaccination camp. However he was right up to a point. The drainage system was certainly in need of proper organisation and better construction. We can now look back and see that the system of waterworks and sewage disposal set up in Britain by our public spirited Victorian ancestors in the 1880s and 1890s was a truly beneficient legacy. At the same time he was hopelessly wrong about his prediction for vaccination. What would he have said if he could have known that a hundred years after he made his tirade, smallpox was eliminated from the entire world through a combination of vaccination and public health measures? Even so, the whole question of vaccination still comes in for argument—especially in regard to cases of vaccination-related brain damage in young children. A point of interest is that while Mr Wheeler was arguing against *compulsory* vaccination, he was doing so on grounds of it being a 'useless and injurious rite'. He was not arguing (as has been the case in modern times in Britain in relation to the fluoridation of drinking water to improve the nation's teeth) on grounds of the infringement by the government of individual liberty and freedom to choose.

Yet another correspondent, a Mr William Godden, writing in 1898 again brought up the subject of vaccination in passing, while airing his views about the sources of disease. He blamed the unfortunate victims of illness for their own condition. His opening remarks were attacking another correspondent's advocacy of flushing the colon as a means of keeping fit:

... What appears to me a very great objection to the operation [i.e. flushing the colon] is that it is altogether unnatural; and this is what I have always felt with that disgusting operation vaccination. If either of these performances is really necessary for the elimination or prevention of disease, one is almost driven to think that Nature, or the Creator—whichever you please—has sadly bungled, and that the man who wrote 'God never left His work for man to mend' scribbled a lie. But, I don't believe in such things and I'll tell you why. ... I can truthfully say I have never been seriously ill, and I have never patronised a doctor.

Nevertheless, for by far the greater part of [my] thirty-nine years I have been closely engaged in arduous business pursuits, and my 'days' have always more closely approximated to sixteen hours each than to eight. Many a time I have been detained at my place of business until midnight, after which I have had to walk about five miles to my home. And then, if the early mornings were starlit or moonlit, the temptation to put in half-an-hour with my 3 inch telescope was too strong to resist. Yet I feel as fresh tonight as I felt twenty years ago. The time is 8.30 pm; but if I were persuaded that any good would come of it, I would no more mind starting to walk to St. Albans tonight (distance $18\frac{1}{2}$ miles) than I shall going to bed presently. And, ... I have eaten (some) meat, and drank (some) wine &c. in the interval. So that, speaking from experience, it seems to me that disease arises and develops from four sources:—Laziness, dirtiness, drunkenness, and gluttony. I don't mean to suggest that everybody in bad health is either lazy, dirty, thirsty, or a big eater; but I do declare my belief that industry, cleanliness and temperance are better preventives of small-pox, fever, and other abominable conditions, than are vaccination or flushing the colon . . . .

Mr Godden also went on to say that in his view the educational value of the *English Mechanic* was far greater than what went on in schools, and that adequate discussion of such subjects in its columns would improve the health of the community more than would 'whole tons of pills'.

Tobacco and its effects have long been a source of controversy in medical circles. At one time smoking was recommended as a means of preventing diseases of the lungs. It was known, however, that there were unpleasant 'side effects' although these were often associated with smoking to excess rather than with smoking as such. A most unusual use of tobacco was reported in the *Scientific American* in 1869, relating to an incident during the Civil War in 1864 where tobacco was claimed to have been instrumental in 'curing' a case of tetanus in a wounded soldier:

A soldier who had been wounded in one of the 1864 campaigns had had lockjaw for forty hours, when the major of his regiment, as a last resource, cut a piece of navy tobacco (about three inches square) put it in a mess pan with boiling water until it was hot through, and saturated with the water; taking it out he allowed it to cool so as not to blister, then flattening it out, he placed it on the pit of the man's stomach. In about five minutes the patient turned white about the lips, which also began to twitch—the man was getting very sick—and in nine or ten minutes the rigid muscles relaxed and his jaws fell open. Indeed it seemed as if the patient would fall apart and go to pieces, so utterly was his entire muscular system relaxed. The tobacco was immediately removed and some whiskey gruel given to stimulate him. Next day the man was taken along in ambulance, and in a few days mounted his horse all right.

Also in 1869, a correspondent to the *English Mechanic* enquired about a means of steadying his hands which became shaky after smoking. This raised several responses, including one from a 'Tobacco Manufacturer' who not surprisingly warned the enquirer not to give up smoking, as the consequences of that might be worse than continuing in moderation.

The best and only antidote I know for the effects of tobacco on the nervous system is tincture of ferric chloride, or generally known by druggists as

Smoking and heart disease

tinct. ferri sesquichloride, from 10 to 20 minims (two drops are about one minim of a tincture) in a glass of sherry—or in water if alcoholic liquor is objected to—taken immediately after smoking, or other excessive use of tobacco, is a complete restorative. If you were to take this two or three times a day your shakiness would probably much diminish; but let me give you one piece of advice, not to give up smoking, as at your age it would do no good; but don't smoke on an empty stomach. After a meal as much as you please.

Another interesting response dealt solely with the symptoms rather than the cause of the shaky hands. The solution offered was a mechanical gadget that would be strapped to the hands so that they would be steady enough for the afflicted person to write properly. In effect this consisted of a leather strap which could be buckled round the hand. The strap contained a pocket which held one end of a brass rod. The outer end of the rod carried a lead weight rather more than an inch in diameter, this being supposed to exert a sufficiently steadying effect. An additional support could be had by holding the weight by the other hand.

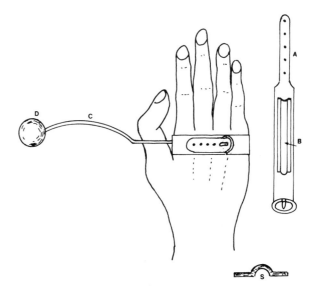

Surprisingly as long ago as 1898, a connection had already been made between smoking and heart disease, yet it has taken until now for this effectively to be brought home to the general public. This article is taken from the *Family Doctor*.

Smoking as a rule agrees with persons for many years—perhaps for twenty years and longer—although by degrees cigars of a finer flavour are chosen.

But all at once, without any assignable cause, troubles are experienced with the heart, which rapidly increase, and compel the sufferer to call in the help of the medical man. It is strange that persons consuming cigars of ordinary quality, even if they smoke them very largely, rarely are attacked in that way. The excessive use of cigarettes has not been known to give rise to similar troubles, although it is the cause of complaints of a different nature. The age at which disturbances of the heart become pronounced varies very much. It is but rare that patients are under thirty years of age; they are mostly between forty and sixty years old. Persons who are able to smoke full-flavoured Havanas continue to do so up to their death.

If we look around among the better class of society, who, it is well known are the principal consumers of such cigars, it is astonishing to find how many persons with advancing years discontinue smoking. As a rule affection of the heart has caused them to abjure the weed. In such cases the patient has found the best cure without consulting the medical man. If he makes up his mind to discontinue smoking at once, the complaint frequently ceases at once; in other instances it takes some time before the action of the heart is restored to its normal state. In such cases, besides discontinuing smoking, relief must be sought by regulating the diet, taking only easy digestible food, light beer and wine in moderate quantities, abjuring coffee, as well as by taking short walks, residence among mountains of moderate elevation, and suitable interior treatment. By taking this course all symptoms disappear in the course of a year and do not reappear if the patient does not recommence smoking. In a third category of cases the more acute disturbances leave the patient; he feels well and hearty; but an irregularity of the heart, more or less pronounced is left behind.

With the benefit of hindsight we can see that the essentials are there, yet it has taken the best part of a century for hard evidence to be accumulated and for the message to be convincingly put across to the general public. There are, however, two things which stand out in this article. Firstly there is the implication that it is only those of upper class 'genteel breeding' who are delicate enough to be vulnerable to this affliction. Those of the lower classes were clearly assumed to be as tough as old boots and immune. The second point is the lack of evidence in connection with cigarette smoking. Presumably again this was because cigarettes, being very cheap, were smoked mainly by the lower classes and were not worth investigating.

One unavoidable affliction that all classes suffer is of course old age. In 1900 the physicist Nicola Tesla came up with an invention that was claimed to reduce the ravages of advancing years.

Nicola Tesla has made a somewhat amusing contribution to medicine. He has found a way of warding off the ravages of time from the surface of the body. This is of special interest to the ladies. He tells us that between 4,000 and 7,000 microbes fall on every square foot of the human body and settle there in 24 hours. If we could see the surface of the body with a microscope we would see it swarming with millions of germs. This would not only, he says, be a hideous sight (he evidently pictures them in his mind's eye with heads and long claws) but they would be seen to be eating the skin and destroying its freshness at a rapid rate. The reason that old people are yellow and wrinkled is because the microbes have for years fed upon their skins. Tesla recommends in the first place thorough washing of the skin once a day with alcohol, and has invented a battery which shoots the microbes off into space with great violence, sometimes to the distance of four or five feet. If this discovery of Tesla's leads people to wash who would otherwise not do so, it would be a boon indeed. We are afraid it will appeal more to those over-sensitive people who are already tottering under the weight of many other more or less senseless fads.

According to the *Medical Press* a previously unsuspected source of illness is riding on public transport, not so much by coming into contact with other people, washed or unwashed, as by getting cold shins on the draughty exposed top decks of omnibuses without adequate protection for the legs. This report dates from 1896.

The 'cold spots' meaning thereby the surface areas peculiarly susceptible to cold, are principally the nape of the neck and the lower part of the back of the head, the front of the abdomen and the shins. The acute discomfort and the sense of impending disaster which results from the steady play of a current of cold air upon the neck from behind are well known. The necessity of keeping the abdomen warmly clad is also generally recognised, though perhaps not as generally carried into practice.

Curiously enough, few people are conscious of the danger they run by exposing the usually inadequately protected shins to currents of cold air. This is the usual way in which colds are caught on omnibuses. When driving, one takes care to cover the legs with a rug or waterproof; but on the more democratic conveyance rugs are not often available, and the reckless passenger by-and-by awakens to the fact that the iron has entered into his soul—in other words, that he has 'caught cold'.

People who wear stockings, such as Highlanders, golfers, and cyclists, invariably take the precaution of turning the thick woollen material down over the shins, the better to protect them against loss of heat, though incidentally, the artificial embellishment of the calves may not be altogether foreign to the manoevre. This is an instance of how all things work together for good. It does not of course follow, because certain areas are peculiarly susceptible to cold, that a chill may not be conveyed to the nervous system from other points. Prolonged sitting on a stone or even on the damp grass is well known to be a fertile source of disease; and wet cold feet are also, with reason, credited with paving the way to an early grave.

By way of explanation, a point to note is that the term 'driving' refers to travelling in one's own private carriage—clearly again we have the distinction between those for whom the article was intended and the lower classes. What is surprising is that the thought of 'carriage folk' so recklessly travelling by public transport was entertained at all. Perhaps the writer of the article had in mind the 'city gent' travelling between his office and the railway station.

One of the complaints that the so-called higher classes probably did not suffer from was varicose veins. If the correspondence on the subject in the *English Mechanic* is anything to go by, varicose veins must have been an affliction of almost epidemic proportions. Perhaps the condition was worsened by long hours at work, many of which were probably spent standing up. Cycling came in for some criticism as aggravating the condition, and at the opposite extreme 'walking in wet grass' was claimed to be beneficial. Amongst all the 'cures' were elastic bandages obtained through *Exchange and Mart*, or just a change of diet 'to prevent the blood becoming too thick'. One rather athletic treatment recommended in 1900 was as follows:

Get a bandage of 'shirting calico' of $2\frac{1}{2}$ inches wide, and beginning at the ankle, bind it spirally round the leg, or take it by the middle and place it on the ankle in front, then cross it at the back, and then at the front as you bind it and working upwards. It must be tight. This requires dexterity. If the bandage is applied while the foot is as high as the shoulder, it would be better, as the veins would be emptied partly by this position. A friend of mine tried the elastic stocking first, and finding it useless, used the calico bandage for 50 years. If neglected the over-distended vein forms at last an ulcer, which after a time may burst, and the whole of the blood in the body will be lost in three minutes, and death ensue.

How the foot was to be raised 'as high as the shoulder' is difficult to imagine unless the sufferer was lying down. He (or she) wouldn't then be able to apply the bandage without assistance. Alternatively, standing up on one leg whilst applying the bandage calls for considerable gymnastic skill unless support is available. A rather more 'scientific' method of treatment might be to turn once more to electricity as a correspondent by the name of Jno. Rogers proposed:

The best course to pursue is electric treatment. If any of your friends possess a battery, they would doubtless lend it to you; or if you have one so much the better. Apply the positive secondary current a little below the varicose parts, with negative a little above, and pass them up the limb a short distance from each other. The veins being relaxed require the contracting influence of the positive. It is well, in addition to apply bandages or laced stockings, and to occupy a recumbent position, avoiding exercise as much as possible, and bathe the limbs in cold water. . . .

One could be forgiven for wondering what is 'positive secondary current' and why the positive causes contraction. Presumably Mr Rogers was trying to say the direction of current was important, although it is not clear whether he thought positive current emerged from one terminal and negative from the other, and why 'secondary'. Did he imply a secondary (i.e. rechargeable) battery, or did he perhaps have in mind some form of induction coil operated from the battery? In any event he was playing safe by advocating both electrical treatment *and* a supporting bandage.

Whilst on the subject of legs it might be appropriate to consider the afflictions of the feet too. Here is an 1869 treatment for corns that is definitely not for the squeamish:

Never cut your corns, but pull out the eye with your thumb nail, holding your toe tightly with finger and thumb to deaden the pain, of which don't be too timid; then apply caustic in the hole left and on the hardened skin around; and when this appears to have acted well (from 6 to 8 or 9 days) then pull it off. But repeating this process two or three times you will be sure to be relieved, if not quite cured. Persevere and you will succeed. The writer has lost three by these means.

Another common problem is 'ingrowing toe nail' but it is doubtful whether anyone today would be willing to follow the treatment advocated by someone describing himself as a 'Rangoonite' in 1891.

I occasionally suffer from this distressing complaint, and I get immediate relief by adopting the following remedy. At the first sensation of pain I scrape the nail as thin as possible and when retiring to rest I apply a bread poultice to the toe—which is easy enough with a little gumption—and over this I draw a dirty sock. In the morning, as a rule, I am quite relieved. Should any symptom of pain still remain, I put a poultice on during the day, when my duties will allow me to do so, and another one at night if necessary. This is a simple remedy and very efficacious.

Problems with feet of course frequently stem from poorly fitting shoes—not to mention poorly designed ones that are more in keeping with fads and fashion than with sensible 'scientific' principles of design.

162   *The shape of feet*

In 1898 a Dr Royal Whitman in the New York paper *Popular Science News* had this to say on the subject:

> ... the object of the shoe is to cover and protect the foot, not to deform it or to cause discomfort; therefore, the one should correspond to the shape of the other. If the feet are placed side by side, the outline and the imprint of the soles will correspond to the accompanying diagram.

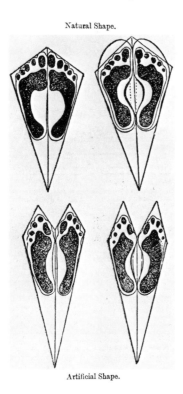

    The outline demonstrates the actual size and shape of the apposed feet, emphasised by inclosing them in straight lines. Thus each foot appears to be somewhat triangular, being broad at the front and narrow at the heel. The imprint shows the area of bearing surface, and owing to the fact that but a small portion of the arched part of the foot rests upon the ground, it appears to be markedly twisted inwards. The sole of the shoe, if it is to inclose and support the bearing surface, must also appear to be twisted inward in an exaggerated right or left pattern; it will be straight along the inner border to follow the normal line of the great toe, and a wide outward sweep will be necessary in order to include the outline and thus to avoid compression of the outer border of the foot.
    I have found this statement of a self-evident fact, and the demonstration of the true form of the foot to be an almost indispensible preliminary to

An artificial foot

an intelligent discussion of the relative merits of shoes, and indeed somewhat of a revelation to those who have thought of the foot only as it has been subordinated to the standard of the shoemaker.

The shape of the foot was important to makers of artificial legs, and it is quite surprising to find what an extensive and thriving trade there was in all kinds of artificial limbs. Many people were not beyond making their own. In the days before 'safety at work' became a political issue and before Her Majesty's Inspectors of Factories were established as part of the scene in British industry, accidents were rife. They were seen almost as only to be expected and just a normal part of life—to be accepted with resignation as one was supposed to accept one's 'God given' place in society. It is particularly sad to see how many young women and children were disabled through tragic accidents at work. What makes it worse is that the victims rarely, if ever, received any form of compensation or redress and what is more they somehow had to fight their disabilities and return to work simply in order to survive.

A rather unusual artificial foot was described in *The Lancet* in 1899. It was claimed to be the invention of Henry Yearsley, an engine driver living at 6 Gleaves Road, Eccles, Lancashire. The most unusual feature about the device was that it was pneumatic. It had to be inflated with a bicycle pump. What might happen if the wearer accidentally stepped on a drawing pin and had a puncture is anybody's guess, but according to the report the idea worked well.

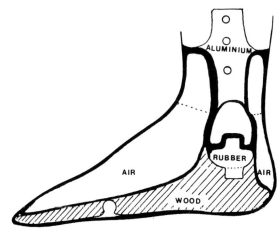

Yearsley's artificial foot, 1899

This is an artificial leg and foot intended for use in cases of amputation below the knee-joint. The chief novelty consists in the foot being mainly a hollow rubber chamber which is inflated in exactly the same way as is a bicycle tyre. The skeleton of the foot, so to speak, is of wood, and contains within it a rubber-faced joint which permits movements like those which take place at the ankle.

The illustration shows very clearly the construction of the foot. A pair of rubber pneumatic pads surround the stump itself, so that no undue pressure is exerted on the tissues. We think this form of artificial leg is likely to prove very useful in suitable cases, as it is light and of sufficient elasticity. We have seen a patient who has undergone amputation of both legs below the knee, and who was wearing two of these limbs. Although having little or no command over the knee joint, he was yet able to walk very fairly well, and to go up and down stairs safely.

The next example is an artificial arm and hand made solely to serve an industrial function with no concessions to appearance at all. It was intended to enable a sewing machine operator to put tacking stitches into her fabric by hand. It dates from 1870.

I have just invented a very useful mechanical contrivance, a photo of which I enclose. Three weeks ago I had a machinist apply for an arm amputated a few inches below the shoulder, and she wanted, amongst a number of appliances, an instrument that would hold the work like the natural fingers, to enable her to hem, stitch and tack her work for the machine.

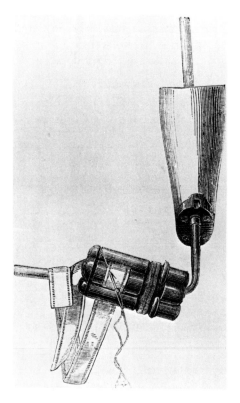

Artificial arm and hand, 1870

After many trials I succeeded, and so successful is the appliance, and simple, that in a short time she was enabled to stitch and hold the finest or coarsest fabric with almost as much ease as with the natural fingers.

The instrument is fixed in the wrist, and is removed instantly on touching the spring. There are three rods corresponding to the thumb and the first and second fingers; the two under rods are fixed; the top one, over which the work is drawn, is movable in any direction and bound to the other two by a vulcanised rubber band; this band according to its tension keeps the work. On to each rod is a piece of brass tube, which revolves when the work is drawn. Over the two tubes is slid a piece of vulcanised rubber pipe. The work is placed in the instrument precisely as it is over the index finger, the vulcanised rubber holds the work just like the soft cushions of the fingers; as the person stitches she can draw the work, the rollers revolve, and thus they can hem and stitch at any length. Ladies who have lost an arm at the shoulder, above or below the elbow, may now by the use of this instrument follow their favourite employment which through their loss they may have long been deprived of.

It is hardly likely that any 'lady' as defined in Victorian terms would have been in that unfortunate position. Neither would a lady have wanted her fashionable friends to see such an appliance. One must, however, admire the industriousness of the inventor, having accomplished his objective within three weeks—invented, constructed and operational, after having carried out 'many trials'.

Surgeons in the nineteenth century became extremely proud of their skill and *speed* in carrying out amputations. Without proper anaesthetics speed was absolutely essential. Using hand operated instruments and handsaws meant the surgeon had to be a skilful operator. Despite Faraday's work on electromagnetism and the subsequent development of the dynamo and the electric motor it was not until the end of the nineteenth century that anyone had the idea of using an electric motor to drive a circular saw and adapting that for surgery. Indeed it was only after the Great Exhibition in Paris in 1881 that electricity was seen to have tremendous application as a more convenient prime mover than the old steam engine. This report dates from 1900:

An electrically-driven saw has been found to be of great use in surgery. The shaft upon which it runs is connected with a motor by a flexible spiral coil inclosed in a braided sheath. The machine has been extensively used in the larger hospitals, and the operations that have usually been fatal with the old hand-saw have been very successful with the new ones.

In 1866 a Frenchman by the name of Blanchet devised a kind of 'artificial eye' that he claimed could restore sight to a damaged eye. The gadget, which he called a 'phosphore', was in effect an optical replacement for the cornea. It would of course only have been able to restore very limited vision and even then would have been quite useless unless the retina was still functioning properly. It was described as consisting of an enamel shell and a tube closed at both ends by glass lenses. These lenses would have had to be made to suit individual circumstances. This 'phosphore' would be inserted into the front of the natural eye. The actual operation to do this was described as follows:

The patient's head being supported by an assistant, the upper eyelid is raised by an elevator, and the lower one is depressed. The operator then punctures the eye with a narrow bistoury [a long narrow surgical knife with a fine sharp point] adapting the width of his incision to the diameter of the phosphore tube which he intends to insert. The translucent humour having escaped, the phosphore apparatus is applied, and almost immediately, or after a short time, the patient is partially restored to sight. Before introducing the apparatus it is necessary to calculate the anterio-posterior diameter of the eye, and if the lens has cataract it must be removed. Inasmuch as the range of vision depends on the quantity of the humour left behind, M. Blanchet recommends the employment of spectacles of various kinds.

It was claimed that 'The operation in most instances causes little pain, and when the globe of the eye has undergone degeneration there is no pain at all'. Although there was no explanation of how the 'phosphore' was held in place, the reference to an enamel shell suggests that this was pressed up to the front of the eyeball, giving it a metallic false front, with the tube containing the two glass lenses mounted in the middle.

The reference to 'spectacles of various kinds' refers to the patient's inability to adjust the focus—especially if the natural eye lens had been removed—to accommodate near objects as in reading, and distant objects as in normal walking about. Bifocal spectacles probably would not have been as much help in this instance as would separate spectacles for the two kinds of circumstances, since the movement of the eyeball must have been restricted by the enamel shell on the front making it difficult to move the line of vision through the two parts of a bifocal lens.

Interestingly enough bifocal lenses of a kind had been around a long time. They are usually associated with Benjamin Franklin in the eighteenth century. A rather neat type of bifocal lens was put

Bifocal lenses

on the market in 1899 and described in *Popular Science* as the invention of a Mr Borsch of New York.

For persons whose vision is of such a character that a lens that enables them to see distant objects does not enable them to see print or objects at close range, a bifocal lens is a decided convenience over the old practice of wearing two pairs of spectacles, as one sees portrayed in the portraits of Patrick Henry, one pair resting on his nose and the other pushed high upon his forehead.

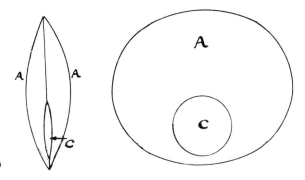

Bifocal lens, 1899

Hitherto the bifocal lens has been made of two pieces of glass of different curvatures cemented together or confined within a frame. The line of junction of the two lenses is annoying to the sight; dirt collects in it, and it is an obstruction to the convenient wiping of the glass. The figure shows the new bifocal lens, which is said to have attracted the attention of the Franklin Society. It consists of a minor short-focus lens C for viewing near objects, inclosed wholly within the interior of a longer focus or less powerful lens A, for viewing objects at a distance. The difference in magnifying power of the two lenses is obtained wholly by the difference in refractive index of the different kinds of glass of which they are made.

At first glance this lens seems to offer tremendous advantage over the normal type of bifocal lens. However, it is actually made in three parts cemented together and the two outer parts would have to be very accurately ground to fit and enclose the inner lens properly. In addition, the inner lens would have to be made of much denser and more expensive glass than the outer lens, otherwise the optical properties of the inner lens would be effectively lost because unless there was a significant difference in the refractive indices of the two lenses, the little one would effectively become an indistinguishable part of the outer one and would not materially alter the focal length. The cost of production as well as materials would probably mean lenses of that kind would have been very expensive. What is more,

Still on the subject of eyes, a school Medical Officer named Dr Brudenell Carter in 1896 advocated the introduction of 'seeing competitions' in schools so as to improve the nation's vision.

Dr Brudenell Carter wishes to introduce a new game in schools, and one of a novel character. He has been examining for the Education Department, the eyesight of elementary school children, and, as a result, he thinks that 'seeing competitions' would be a good thing for improving visual powers. It is pleasant to find that he has no doleful state of matters to disclose; the children are not 'with blinded eyesight poring over miserable books', their eyes indeed are not injured in any way by the conditions of school life. Dr Carter has tested over 8,000 children in the London schools. About 40 percent had absolutely normal vision in both eyes, and an equal number had subnormal vision in both, the remainder being perfect in one eye only, which, in the majority of cases was the right. Town children have not such good eyesight as their country cousins, and girls are not so well endowed, on the average, as boys.

Hypermetropia or 'long sight' is the most frequent irregularity, and, unless aggravated by poor feeding or bad light, it is only a matter of early resort to spectacles. But when the subject is young this defect can be got over by the aid of the elastic and adaptable condition of the organs. Many would have better sight if they had been trained as children in the art of seeing, and Dr Carter urges very strongly the encouragement of this branch of physical education. Rapid and clear observation is to a great extent a matter of teaching and practice, and in the case of animals and of *savage races* [my italics] who have to place so much dependence on their natural powers, it is very highly developed. There is not the least doubt that civilised races could enjoy much greater usefulness from their eyes with proper attention to training . . .

For those who were considerably overweight, a rather drastic way of getting rid of unnecessary fat was described in 1900. The subject of this article, it was claimed, lost 42 pounds in weight in six months and reduced his waist measurement by 11 inches.

Becoming a total abstainer, he rose at six o'clock each morning, and after drinking half a pint of hot water, walked four miles, on returning had a quarter-hour's 'Sandow' and a cold bath. Breakfast consisted of three slices dry brown bread toast, pinch of salt, and one cup of weak tea, no sugar and little milk; not another drain of drink until tea. For dinner, lean meat, brown bread, and vegetables, except potatoes, carrots, parsnips, beetroot, beans and peas, few eggs, little cheese, no pies or pastry; in fact no starchy or sugary foods whatever. Fruit was eaten when thirsty; for tea the same as breakfast, no supper, and before going to bed one pint of hot water with juice of half a lemon in it. Stiff, isn't it? Anyway, by Sunday (he started on the Monday previous) he had lost 8 pounds in weight, and felt nearly dead so had full diet for the day, and returned to system on the following day, and did not depart from it again during the rest of the six months, with the result aforementioned.

Another writer on the same theme three years earlier (1897) had this to say:

Most fat people are primarily healthy people, with good appetites, good digestions, and cheerful dispositions. They are also usually fond of rich food and sweets, and will not work or walk more than they can help. Usually too, they are long and sound sleepers. Exercise, unless pushed to very great lengths, will not reduce a healthy fat person. His appetite increases and he eats so much more that he may actually gain weight, whilst taking an enormous amount of exercise. The first thing is to diminish the intake. Less must be eaten at every meal. The sufferer from obesity must always get stout whilst increasing his amount of outdoor exercise. Sweets of all kinds must be avoided. Lean meat, white fish, salads, poor vegetables such as boiled celery, spinach, cabbage and seakale must form the dietary. Fat, butter, cream &c. are to be avoided. To descend from the general to the particular; rise at 7 a.m.; breakfast at 7.30: lean chop or steak, cold lean beef or mutton, rabbit or fowl with a little old bread. Tea or coffee without sugar, but may have a little skimmed milk. May sweeten with saccharin or sucrol; no butter. Dinner, 1 p.m.: lean meat, mutton, beef, game, hare, rabbit, fowl and white fish. A small allowance of potatoe, salad and cooked vegetables, cabbage, cauliflower, spinach, celery, seakale. A little boiled rice and stewed fruit after. Tea, 5 p.m. much same as breakfast. Supper: cup of cocoa, Cadbury's essence, Schweitzer's cocoatina, made with water and a little milk. Saccharin to sweeten or sucrol.

As regards medicine, take half a teaspoonful of Tinct. Fuci vesiculosi three times a day in half a wineglassful of water. Avoid alcohol in all forms, above all sweet-made drinks such as lemonade, ginger beer, ginger ale, &c. Take at least once a week—say every Sunday—2 oz. of Hunyadi water with two-thirds of a tumblerful of warm water first thing in the morning on arising. If troubled with constipation take rather less. If conscientiously carried out this system will produce a rapid decrease of bulk, and an increase of strength. Rigid vegetarianism always makes people thin, but it also makes them weak and may completely unsex them.

Well that is a rather stiff warning about vegetarian diets. An interesting point about this account, however, is the availability of substitute sweeteners and saccharin in particular in 1897 when the piece was written. A further warning was to guard against damp feet since that, in conjunction with 'errors of diet' could cause colic of the bowels! However, for anyone suffering from stomach upsets, diagnostic help was at hand in 1898 in the shape of an amazing piece of equipment capable of taking internal photographs of the lining of the stomach. This new apparatus was the invention of a Dr E O Schaaf of 217 South Orange Avenue, Newark, New Jersey who was at that time a recognised authority on diseases of the stomach. The original experiments were carried out with dogs and as soon as the technique was perfected it was tried out on human patients. This is Dr Schaaf's own account of how it was done:

It was while performing a surgical operation called exploratory laparotomy that the idea of photographing the interior of the stomach occurred to me. That hazardous operation would have been unnecessary if the interior of the stomach could have been seen by a photographic process—a safer, simpler, and more scientific method of diagnosis.

Abdominal section is often performed for cancer, the increase in the mortality from which is alarming, for nothing at all approaching it is furnished by the statistics of diseases of any other organ.

A purpose of the stomach camera is to set aside the usual conjecture and liability of error in diagnosing various diseases of this organ, especially in their incipiency.

Last year, after numerous attempts, I successfully photographed the interior of the stomach in a living person, obtaining a picture of the pyloric mucous membrane. Since then improvements have been made in the camera, lessening the difficulty of manipulation.

One of the early difficulties was to properly focus the membrane, and the fault I believed lay with the movements of the stomach during respiration; but this difficulty was overcome by the use of improved films, a lens of higher power, and stronger light, so that a patient need not suspend respiration as before during exposure of the film to a flash of electric light in the stomach. There is no pain connected with the procedure, the sensation being no different from that familiar to patients who wash out their stomach with a soft rubber tube.

The microscopic as well as chemical examination of the mucous membrane and secretions of the stomach, so much relied upon during the past few years in making a differential diagnosis between benign and malignant affections of this organ, is not always reliable, although it will sometimes yield valuable information. Photographing the stomach is a method that is perhaps destined to replace the microscope in the matter of diagnosing stomach diseases. It is more reliable than the examination of tissues whose dimensions are the fractional part of a millimetre.

The camera is of very simple construction, consisting of a water-tight silver cylinder with a lens at one end and a film at the other which will accommodate itself to different focal distances. In order to photograph the stomach, the patient must fast from four to six hours; the mucous and undigested food are then washed out so that nothing but membrane will be in focus when the camera is introduced. About a pint of water is left in the stomach; this dissolves the gelatine capsule that surrounds the lens, and facilitates focussing by keeping the stomach walls apart.

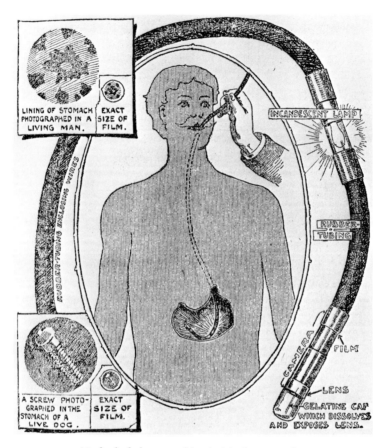

Method of photographing inside the stomach, 1898

The electric light may first be introduced in order that an outline of the stomach be made externally on the abdomen, and the required length of tubing from teeth to focal point be determined. The films are about three-eighths of an inch in diameter, and the picture secured may be enlarged to a diameter of three feet by a projectoscope.

Photographing the stomach is superseding gastroscopy, which had its origin with Mikuliez in Vienna, who in 1881 constructed a gastroscope of stiff metal tubing which he inserted into the stomach. This enabled him to see the cardia. The method caused great discomfort to the patient, and was attended with much risk. To Einhorn belongs the credit of making trans-illumination of the stomach a practical method. I have busied myself with this instrument for several years and the photographic stomach camera is the result of my investigations.

There doesn't appear to have been any way of positioning the camera when it was inside the stomach, other than by pulling on the tube and trusting that it will have oriented itself in a favourable direction. The camera itself was extremely basic and had no aperture stop or shutter. Flashing the light was in effect equivalent to opening what would normally be the shutter. Presumably too, only one picture at a time could be taken since the apparatus would have had to be withdrawn to change the film. The arrangement of the light was extremely ingenious. It had its own little compartment inside a glass cylinder at just the right distance from the camera for it to illuminate the inside of the stomach without flashing straight into the camera directly.

It couldn't have been a particularly pleasant experience for the patient to first of all have to swallow an electric light so the doctor could determine the length of camera tube to insert, then to have that withdrawn and swallow the camera. If several photographs had to be taken the patient would have to go through the whole process of swallowing and bringing up the apparatus every time. Unpleasant though that might have been, it was probably nothing to coping with the fearsome 'gastroscope'.

After the discovery of x-rays in 1895, the whole business of diagnosing internal problems began to change. As with electricity, x-rays were hailed as a new solution to every problem, until things began to go wrong and the consequences of over exposure were realised. The x-ray photograph illustrated here is breath-taking, not only in the extent of coverage of the patient's entire body, but also in the enormously long exposure times.

The photograph is actually a composite. Nine plates in all were exposed. It was carried out at the Case School of Applied Science, Cleveland, Ohio, in 1897. The exposure times ranged from *three minutes* to as much as *twenty minutes*. The knee required 8 minutes, whereas the pelvic region was exposed for a full twenty minutes. The patient was 31 years old. It would have been interesting to know what happened to him afterwards. An even more interesting question is, how on earth did he keep still for twenty minutes.

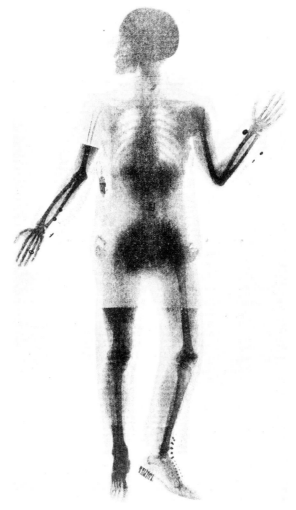

X-ray photograph, 1897

There would have been no problem in keeping still for the lady who was the subject of the next item. This describes an apparatus for correcting stooping shoulders in young women:

Let the young lady be properly measured for a good strong well-steeled V-boned pair of stays having a broad wooden busk, say $1\frac{1}{2}$ inches wide and 16 inches long, with two good side steels about the same size coming well up under the arms and back. Then get a strong tape $1\frac{1}{4}$ inches broad and of sufficient length when it is crossed at the back to come from the throat to the stomach.

Now supposing both are ready, let the wearer place the endless tape over the stomach, take the other end crossed over the back, and pass it over her head so that it comes round the throat. Then lace up the stays tightly, or the front tape will tend to pull them upwards when laced. The tape being in position round the throat, the portion going round the lady is slipped under the inverted hook which is riveted on the bottom of the busk. To get the tape under this hook, the head for a moment should be well thrown back. It will now be seen that it is impossible to stoop forward in the waist, because the wood busk will not yield, and if the tape is rightly adjusted, the neck and shoulders cannot come forward without causing very unpleasant pressure on the stomach. Any leaning forward causes the tape to pull tighter under the hook and to relieve this pressure the wearer naturally straightens herself up. By some it is considered irksome at first—this impossible forward movement or stooping; but after a time, it is said to be so delightfully supporting that the wearer cannot be persuaded to do without them. But I always insist upon my daughters (clerks to a railway company) leaving them off altogether on Sunday, or on other days when not going to the office; but on no account would I let them wear those flimsy things that fasten in the front—sold in the shops. They are, I feel sure, from the flexible nature of the busks, the cause of three parts of them stooping. As to bicycling, they say the tighter laced they are, the more comfortable riding...

One would have to be not only a contortionist but somewhat masochistic to wear such an article only one degree removed from armour plating. An appliance with a similar purpose but rather less severe was the invention of a Mr E J Chance in 1887.

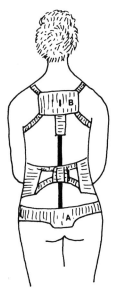

Chance's back support, 1887

A belt A surrounds the pelvis; a single upright rod passes from this belt as high as the shoulders; a pad B movable by a key is attached to the upright rod; the shoulder straps proceed from this pad; an abdominal belt is attached by straps to a pad opposite the lumbar region, and this pad is affixed to the upright rod. The apparatus is so fitted that when the patient is seated the lower portion of the pelvic belt A comes in contact with the chair, so that support in resting the back is complete.

As for the treatment of wounds in the nineteenth century all kinds of potions and concoctions were thought beneficial. A correspondent to the *English Mechanic* in 1887 had this to say:

My object ... is principally to insist on the necessity of avoiding as much as possible the stitching of wounds in the dressing of them. This barbarous mode of treatment should have been superseded long since in the progress of the times. I always consider it a very dangerous treatment for a tender wound, as the stitching must excite more or less powerfully to inflammation. Theory is not to be depended upon at all in medicine for the operations in the living body transcend our powers of theorising. Thus, common salt is a strong antiseptic; therefore, theoretically, salt when put into a wound ought to have a beneficial and soothing effect, instead of having, as it does, an injurious and very irritating effect. Considering the antiseptic power of the mild boracic acid, we should expect that the liquid impure carbolic acid, that black venomous divinity of the present time, ought soon to become a broken and fallen idol. It is to the antiseptic action of the blood alone that the cure should be left as much as possible; and accordingly the blood's curative action should be assisted by inducing easy circulation in the sore flesh with emollient ointments. I agree with one correspondent that cold water should be avoided, for though cold itself is beneficial, yet I look upon the astringent principle of water-dressing as not only a humbug, but as to some extent injurious. Astringents are of the opposite nature to emollients, and if emollient spermaceti ointment is found so serviceable in external inflammation, by what reason can it be supposed that astringents, producing the opposite effect, can also be of service in the same cases? The astringent effect impedes the circulation of the blood through the injured flesh, whereas emollients promote its circulation; and which, I would ask, is the better principle according to reason? Cleansing, however, with tepid water and soap lather is often beneficial. I do not agree with another correspondent that sticking plaster should be discarded, for it is often very useful, both on account of the protective mechanical coating it supplies, and by its keeping the edges of a wound close together. I would not think of putting sticking-plaster on an inflamed skin; but I find that nothing is so good for a skin affected with itching. Perhaps, in general, the best soothing application for wounded or inflamed parts is vaseline. This substance is perfectly free from any acrid or irritating quality; it never becomes rancid, contains no oxygen, and has a strong penetrating power which makes it soak into the flesh. For avoiding the unsightly appearance of a sore lip nothing equals vaseline, when early applied, as my experience teaches me, since it cures the sore without giving it the "raw" appearance that the other unguents give it. In all cases I should be against using irritating applications of any description on inflamed surfaces, as I am convinced that they have caused the loss of many a limb, and even many a life.

An Apothecary

DOCTOR STARVES, TURNS QUACK AND THRIVES

A Bordeaux medical journal is responsible for the following amusing, if not instructive anecdote:—A certain individual who called himself Alexis, having been accused of signing prescriptions, besides otherwise trespassing on the prerogatives of the medical fraternity, the police commissary of the district—which is described as one of the most *selected* in the French capital—received instructions to investigate the charge and, should he find cause, arrest the delinquent. The official accordingly proceeded to the house where it was said the medical canons were habitually violated, and a single glance sufficed to satisfy him that he had surprised his quarry *in flagrante delicto*. Magnificiently installed in a suite of sumptuously furnished apartments, with three spacious waiting rooms filled to overflowing by well-dressed and evidently moneyed crowds of patients, the reputed quack submitted to the intrusion of his inauspicious visitor with dignified calmness.

'You will continue to admit callers' he observed to his janitor as he was preparing to accompany the commissary to the police station 'for I shall be back in a few minutes.' The officer did not exactly smile when he heard this remark, for while on duty the smiles of an *agent de police* are as rare as black swans, but the case was as clear as daylight, and a close observer might have detected some indications of incredulous merriment near the corners of his mouth. 'You will do me a very great favour, sir,' said the accused man when they had reached their destination, 'by according me a moment's private audience.' 'For what purpose?' inquired the man of law. 'There is something most important that I wish to communicate to your ear alone,' replied the other. The commissary, for all his apparent sternness, had an obliging disposition. 'Clear the room' he exclaimed, and the pair were left free from witnesses. Upon this M. Alexis drew forth his pocket-book, and extracting therefrom a parchment diploma, he spread the document out before the eyes of his astonished interlocutor. 'I beg of you,' he cried in imploring accents, 'not to betray me. I should lose every single patient I have got if they knew I was a properly qualified practitioner. When first I obtained my degree I set up in the Fauberg Saint-Antoine, fixing a brass plate on my door and carrying out all details *secundum artem*, but in six months I had spent all my money and was starving. I consequently removed the plate, suppressed my surname, and established myself yonder as a bone setter. Since then my affairs have gone on wheels, and I am in a fair way of making my fortune. I can advertise freely now, and knowing something about the business have succeeded in avoiding fatal mistakes. I implore you again to be merciful.'

For a little while the commissary felt perplexed, but his doubts did not last long. When a *rebouteur* posed as a medical man it was his duty to pursue him, but the converse did not hold good. There was no law to prevent a practitioner turning bone-setter if he liked; and so M. Alexis, the qualified quack, continues to flourish.

English Mechanic, 1897

AN IDIOT WHO IS A GENIUS

Jephtha Palmer, a white man who lives at Fairmount, in the North Georgia Mountains, is an idiot and yet a genius. In all matters except the construction of machinery and the composition and production of music his mind is impotent, says the *Louisville Courier-Journal*.

Palmer was born in 1848 near his present home. His father was a poor tenant farmer who spent his time toiling on an unproductive farm, and died as he began—ignorant and penniless. Jephtha's early childhood showed no indication that he would ever be more than a hopeless idiot. He could not intelligently call for food or find his way from the farm to the house when he was seven years old, and his parents and all who knew him, supposed that he would have to be attended to all his life as a helpless infant.

But when Palmer was in his eighth year a travelling horse-power wheat thresher came to his father's home, and the operation of the machine so greatly amazed the boy that he stood gazing at it the whole day, much to the amusement of the workmen and bystanders. Finally when suppertime came, his father took him by the arm and led him off to the house against the furious protests of the boy, who remarked as he walked away: 'I am gwine to make me one o' those things.' For several days after the threshing machine left the Palmer home, it was noticed that Jephtha was very busily engaged at something out behind the yard fence, and one day when his mother was passing that way she was astonished to see that he had a complete model of the wheat-threshing machine in successful operation. He had made it with his pocket knife of pine bark, and was using strings for belts.

Shortly after this he went with his mother to the washing place, and while he was stirring the fire under the wash pot he accidentally struck the stick against the top rim of the pot, and the clear ring of the hollow metal filled him with a wonder he had never before known. He struck the pot again and again, and was delighted with the varying sounds. He at once set to work, and in a short time completed an instrument of long and short strips of raw hide stretched across a triangular wooden frame, and on this he was able to make good music. Up to this time he had never seen or heard a musical instrument of any kind whatever.

When sixteen years old Palmer made a clock entirely of wood, and used stone weights. This clock, which was constructed with the rudest of tools, ran for a great many years and kept excellent time. During the next five years he made some forty or fifty of these wooden clocks with stone weights, and, although none is now in use, quite a number can be found in the houses of the neighbours. While still very young, this remarkable person built, without information or suggestion on such matters, a mill on the order of the pounding mill, the power for which he obtained by damming up his father's spring branch. The stroke of this peculiar mill fell on an old door shutter, and the noise of the pounding could be heard for miles over the neighbouring mountain settlement.

When Palmer reached manhood he devoted himself to the repairing of clocks, and for many years was a familiar figure on the country roads and the streets of the nearby towns. He carried the few tools he had in an old basket and went bare-foot in all sorts of weather, usually with his trousers rolled to his knees, and his sleeves elevated to his elbows. When asked why he did not wear shoes, his answer was that they smothered him. He would go on long trips over the rough mountain roads, frequently carrying four or five old clocks and other plunder on his back.

He became known all over the region, and was a sort of pet and privileged character, stopping to spend the night wherever he found himself or his fancy dictated. If his host had a clock or a musical instrument that needed repairing he did the work free of charge, and usually spent half the night playing on the accordion, fiddle, or whatever musical instrument happened to be in the house. He has always played all classes of instruments without any instruction, and apparently without effort.

Since his youth he has been able to immediately write out the music of any tune he may hear produced vocally or instrumentally, and he has produced many a score of beautiful and harmonious compositions. Among his compositions are polkas, waltzes, and marches, pronounced by musical critics to be of the highest class.

He has made a large number of crank organs. One of these consists of a round cylinder of wood, driven full of nails or iron spikes, which, as the cylinder turns, strike a row of steel reeds. These were made of the steel ribs of four old umbrellas, and are called by Palmer, the organ's teeth. This organ plays six tunes, and has been exhibited at nearly all the mountain towns in the northern portion of the State. The instrument and its inventor constituted, for a few days only, one of the most unique sights of the Piedmont Exposition at Atlanta in 1889. This was Palmer's first and only visit to a large city. He became greatly disgusted with the crowds and the noise the first day, and did not cease complaining until he was carried back to his mountain home where he married, and has since remained. He made recently a cabinet organ entire, and the music it produces is said to be excellent.

English Mechanic, 1898

The illustration which we give below sufficiently explains itself. Messrs. Cramer and Co. are the designers of the small portable piano, which, thus slung from a frame across the patient's bed enables the bed-ridden musician to pursue his or her art.

Portable piano for invalids, 1897

We are indebted to Dr Frederick Spicer of Devonshire Street, Portland Place, W., for the sketch which we reproduce, and think it probable that many invalids may like to avail themselves of the opportunity afforded by this little instrument of whiling away weary hours.

The Lancet, 1897

HYSTERICAL BLINDNESS

Cure hysterical blindness by charging electrostatically until sparking occurs.

English Mechanic, 1879

Chapter 5
Velocipedes and Flying Machines

... the motor industry owes as much to enthusiasts of all trades, or no trades, as it does to the professional engineers. In particular it owes much to amateurs.

Anthony Bird
Steam Cars 1770–1970, Cassell, London

THE motorcar is popularly, and erroneously, regarded as a twentieth century phenomenon. As a convenient means of everyday personal transport for millions of people of course it is. As a triumph of science and technology it unarguably belongs to the nineteenth century. In 1885 Carl Benz, a gas engine maker in Mannheim, Germany, built and ran a spindly three-wheeled two-seater car powered by one of his small gas engines suitably modified to burn petroleum spirit (the term 'petrol' had not then come into use). At the same time Gottlieb Daimler produced a high speed internal combustion engine—originally for all kinds of industrial and domestic small-power applications—which he tried first in a crude two-wheeled motorcycle, and then in an adapted horse carriage with the horse and the shafts removed. The year 1885 is thus generally accepted as the start of the vast motorcar industry which has since grown up, making 1985 the motorcar's first centenary.

It would, however, be quite wrong to suppose that no one had driven a self-propelled powered vehicle before Carl Benz did so. The chief engineer to Louis XIII, Salomon de Caus (1576–1626) is often credited with applying steam to power a vehicle. Probably the first vehicle to run with any degree of success under its own power was the gun carriage designed and built by Nicolas Joseph Cugnot in 1769 with support from the French government, using a very compact steam engine—two decades before the French Revolution. This machine is said to have carried four passengers at 3 mph but it could only keep up enough steam pressure for a fifteen minute run, after which it had to stop for a rest until steam could again be raised. In 1770 Cugnot built an improved machine which still exists. It can be seen at the Conservatoire des Arts et Métiers in Paris and is a truly awe-inspiring sight.

The valve mechanism in Cugnot's machine was invented in about 1688 by another Frenchman, Denis Papin. Papin was exiled from France with thousands of other Huguenots when Louis XIV revoked the Edict of Nantes in 1685 and he then came to London where he took lodgings for a time with Sir Robert Boyle (he later became a Fellow of the Royal Society and accepted the Chair of Mathematics at the University of Magdeburg). In about 1707, Papin is said to have built a steam-powered boat which he intended to sail from Cassel to Minder and on to Bremen and then to England where he planned to continue his experiments. However, Papin failed in his attempt and not having sufficient funds to continue he is supposed to have uttered the immortal phrase: 'I am now compelled to put all my plans on the shelf of my chimney.'

Isaac Newton also described a steam-powered carriage designed by him in about 1680.

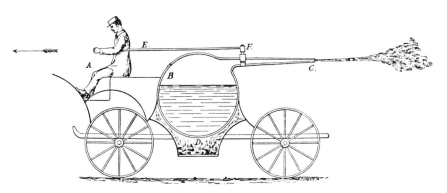

Sir Isaac Newton's steam road locomotive, 1680

Although interesting as an idea, Newton's carriage—if it had been built full-size—would have been a failure. Since it worked by jet propulsion, the mass and velocity of the ejected steam would have had to be many times what is possible in order to move the enormous mass of the carriage and the water in the boiler even at a snail's pace. However a Jesuit priest, Father Verbiest, who was a missionary in China, is recorded as having made a model steam carriage work some time in the 1680s using a kind of geared turbine.

Leaving aside these early experiments and speculations, it may come as a surprise to some to learn that in the first thirty years of the nineteenth century there were literally dozens of steam-powered carriages running in Britain. Unfortunately most of the available investment money went into railway developments and, as is well known, the roads fell into a dreadful state of neglect. Without suitable roads to run on, no vehicles could survive long—and especially so when we consider that the science of metallurgy was almost non-existent. Constant pounding and excessive steam pressure soon took their toll. Engineers received precious little encouragement from the government either. In 1865 the government made matters worse for the horseless carriage fraternity when they lumped together *all* self-propelled road vehicles as 'road locomotives' and required *not fewer than three persons to be in attendance* under the terms of the Locomotives Act. One of the three attendants was required to walk 60 yards in front of the vehicle carrying a red flag during the day time and a red lantern at night. In addition the speed was restricted to 2 mph in towns and 4 mph in open country. Although this Act of Parliament was modified in various ways (especially in 1878 when the legal requirement for the red flag was lifted) it was not until 1896 that it was replaced by the Locomotives on Highways Act and the speed limit was raised to 12 mph. On 14 November 1896 the Earl of Winchilsea tore up a red flag to symbolise the abolition of the old restrictions and every motor vehicle that could be mustered took part in a celebration run from London to Brighton.

One of the things the 1865 Act did was to strangle any experiments with light-weight powered vehicles intended for one or two passengers, since a minimum of three had to be on hand. However, the idea of a horseless carriage was not entirely killed stone dead. Unfortunately space does not allow us to explore all the fascinating developments that took place—often surreptitiously with trials under cover of darkness when the local policeman's back was turned. One important consequence of the Act though, was the rapid development of what were called 'pedomotive' or 'manumotive' carriages. In other words muscle-powered vehicles or 'velocipedes'. Surprisingly perhaps, although primitive forms of two-wheeled 'hobby horses' propelled by sitting astride a cross bar and pushing with the feet on the floor had been around since about 1790, it was not until 1869 that the first bicycles with rotating cranks and pedals were introduced into Britain. These were imported from France where they were constructed by Pierre Michaux together with his two sons Henri and Ernest. The person credited with first applying pedals to a bicycle was a Scottish blacksmith named Kirkpatrick Macmillan who in 1839 built a machine with oscillating treadles connected by rods and cranks to the back wheel. The Michaux arrangement with the cranks attached to the axle of the front wheel was much simpler and more effective.

One of the problems early designers of velocipedes had to overcome was that of steering geometry. Until almost the end of the century, the stabilising effect of having the axis of rotation of the steering forks slightly ahead of the point of contact of the wheel on the road—the castor effect—was not appreciated. An attempt to 'improve' on the normal steering arrangement of the conventional Michaux 'boneshakers' was put on the market in 1870 in the shape of the 'Phantom' bicycle, which was steered from the middle of the frame.

The 'Phantom' bicycle, 1870

A description of the machine was as follows:

The machine is steered from the middle of the frame, or by *both* wheels, and not from the front wheel alone, as in the ordinary bicycle, and as the front wheel only turns half as much as usual, it is never brought into proximity even with the rider's thighs. It is completely railed off too by the frame within which it is enclosed. In turning a corner, or on a curve, each of the wheels are put upon the same arc of a circle; the back wheel, therefore, always *follows* the line of the leading wheel; it passes over exactly the same ground in fact; there is in consequence no drag. In the event of a fall the rider does not get hurt by the machine—that is, he cannot get 'mixed up' or entangled in a painful way between the front wheel and the backbone, as in the ordinary bicycle. The shape of the framework ensures greater speed being obtained than does that of the ordinary pattern. A spring concealed in the steering socket destroys the vibratory concussions usually conveyed to the body. The cranks are readily adjustable to any length of the leg, and when adjusted they do not work loose. Both the wheels have fixed axles, and run in gun-metal bearings fitted with separate oil cups which do not leak. Lastly the wheels are a grand improvement in themselves, being made upon the suspended principle in such a way that they act as springs and assist in dispelling the concussions caused by rough roads. The general result is that a bicycle is produced which is of a decidedly improved appearance, which is much safer to ride, easier to steer, far less dangerous to fall from, and which is much easier to mount and dismount. Finally the *speed*, and consequently the ease of driving are improved by at least a third.

It is interesting to note the emphasis given to 'speed'. A report in *The Times* of 25 November 1869 says that the Liverpool Magistrates defined a bicycle as a 'carriage' within the meaning of the 1865 Act and promptly fined a youth named Carrol for riding one on the footpath. Reports abound of people being fined for 'furious riding' and indeed *The Times* of 24 May 1869 stated that 'We do not think the bicycle is safe for the road'. It sounds very familiar.

One point about the 'Phantom' bicycle which the report doesn't mention is that since the rider's position is almost exactly in line with the steering axis, on turning a corner, he would move sideways relative to the centres of the wheels. In effect, since he would also have to lean inwards to the curve, this would mean that he would have to slightly raise his centre of gravity as he turned, and that would have made turning much harder. It must have been a strange sensation to see the front wheel moving to the side as the handlebar was turned and would not do much to inspire confidence.

Apart from steering problems, almost as soon as the bicycle appeared much brain power was expended in trying to multiply the propulsive effort with less muscular exertion on the part of the rider. One way to double the leg power was by carrying two people. A method of doing this was devised by a John Fielden in 1869.

Fielden's double-driver boneshaker, 1869

The front rider operated in the normal way with his feet on the pedals attached to cranks on the front axle. The rear passenger, however, was expected to contribute both leg power and arm power by means of a pair of detachable levers. The lower end of each lever was fitted with a bowl-shaped stirrup and was attached to a rod connecting with a crank on the corresponding side of the rear axle. It was possible to take the levers off in level country so that the rear passenger could pedal the back wheel directly. What he then held on to to stop himself falling off is anybody's guess. Presumably he grabbed hold of the man in front.

In 1893 a carpenter by the name of A Austin who lived in Stevenage came up with a highly original way of multiplying the propulsive power of a bicycle. He proposed to do away entirely with cranks and pedals and substituted stilts with large rubber feet suspended from the framework in such a way as to effectively lengthen the rider's legs and thus increase the leverage. The rubber feet 'walked' along the road in much the same way as did the rider of the old-fashioned hobby-horse nearly a century earlier. The brakes were applied by putting both feet down simultaneously. Sad to say, Mr Austin didn't stop to think that the arrangement was actually operating at a mechanical *dis*advantage and anyone who tried to ride his machine would have collapsed with exhaustion on anything other than the gentlest of downgrades.

An amazing 'improvement' for the propulsion of 'cycles, motor cars and railway locomotives' was offered in 1897 by Thomas Bennett of 69 High Holborn, London. The normal rear wheel was replaced by a compound wheel driven by gears.

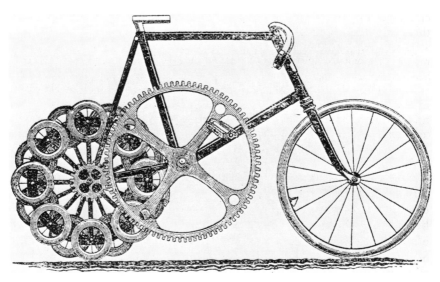

The 'Eureka' driving wheel for bicycles, 1897

A mile a minute may be accomplished with this cycle. This is practically a frictionless machine. . . . The heavier the rider the faster he goes.

This cycle is so much easier to propel than any other because the circumference is divided into 16 parts, and is composed of 16 smaller wheels, only one of which at the same time is in contact with the ground, which offers the resistance necessary for propulsion. When the small wheels come in contact with the ground, the ordinary propelling power obtained from the crank is augmented by the weight of the rider.

The weight of the rider is perpendicular to the centre of the small wheel which is under him. Immediately the toothed wheel is turned ever so little by the larger toothed wheel from the crank, the weight of the rider is lifted off the centre of the small wheel under him: this coupled with the power put forth by the rider on to the pedal causes the smaller wheel under him to, as it were, slip from under him with tremendous velocity (the heavier the rider the greater the velocity) on to the next small wheel.

It is the same slipping from under your motion that a learner on wheel skates feels when the hind wheels of the skate slip foward from under him: his weight goes off the centre and he loses his equilibrium and goes on to the back of his head with great force (the heavier the man, the heavier he falls) but in the present instance, with my cycle, the equilibrium is maintained, and the cycle is propelled forward with the energy thus created. If the skater could have only kept his equilibrium, he would have gone forward as fast as his skate travelled, instead of being left behind on the floor.

This slipping movement occurs 16 times to each revolution of the wheel, and a machine geared up, say, to 150 inches, or five times causes 80 of these small wheels to gain impetus in this way for one revolution of the crank, which will clearly demonstrate the vast speed to be obtained from this cycle. Its value for hill climbing is obvious. Instead of having to turn a large wheel, fresh impetus is gained at each 16th part of a revolution, making the wheel climb a hill with great ease. This is invaluable on that account for motorcars.

For railway locomotives it is proposed to put solid 5 inch steel rollers round the driving wheels of engines against the flange, so that they will revolve upon the metals. This means an enormous saving of fuel, wear and tear to locomotives.

One can only admire Mr Bennett's self-confidence in committing himself to public scrutiny when it takes less than the blink of an eye to see that both his idea and his explanation are unsound. To begin with, the rider would have to have enormously long legs with a radius of rotation of the pedals as big as the illustration suggests. Secondly he could pedal all day and all night without getting anywhere since the little wheels are just idle rollers. He would, however, end up with a very sore bottom from the rear end hopping from roller to roller. On reflection it is just possible that, pedalling forwards in the normal sense, the machine could move slightly backwards because of the reaction on the frame as each little wheel is brought (in an anticlockwise sense because of the gearing) round into contact with the ground. To travel forwards the rider would presumably need to pedal furiously backwards. As for going at sixty miles an hour—Mr Bennett seems to have curiously overlooked the necessity for brakes.

An interesting form of 'motor assistance' for bicycles was thought up in 1869. This is a 'bolt on' windmill attachment.

Windmill velocipede attachment, 1869

A more practical idea was patented in the United States in 1896 by James J Thompson of Jacksonville, Florida. This took the form of an enormous flywheel incorporating a compound system of gears. According to the description which appeared in the *Scientific American* 'By rotating the crankshaft, as in driving the ordinary bicycle, the gears are made also to revolve the flywheel, and power is thus accumulated'. Mr Thompson forgot that the power has to come from somewhere—i.e. the hapless rider—to start with. The rider then has to lug around the extra weight of the flywheel added to the rest of the machine.

Thompson's propelling mechanism for cycles, 1896

Eucycledian bicycle accelerator

A correspondent to the *English Mechanic* in 1879 drew attention to a gadget called the 'Eucycledian Accelerator' and appealed to anyone with practical experience of it to comment on its effectiveness. The device simply consisted of three radial rods attached to the hub of the driving wheel, between the two sets of spokes, with aerodynamically shaped lead weights on the end, giving a similar 'flywheel effect' to the Thompson machine. The writer wanted to know whether the claimed advantages—greater speed on the level and easier hill climbing—more than compensated for the extra weight. This request evoked the following reply:

Having had the opportunity of testing the 'Eucycledian Bicycle Accelerator' I can conscientiously recommend it for the steady and easy-going qualities it imparts to a machine. I have tried it on inclines and find it has marked advantages in facilitating ascent and completely preventing swerving or oscillation in descent. By exerting a back pressure on the pedals I was able to ride slowly down hill with the machine perfectly under control, and this without using the brake.

A slight increase of pressure each successive stroke stores up an immense momentum for future use, and the apparatus is therefore of great service in working against the wind, especially when blowing in gusts. As the whole length of the apparatus (which weighs about 16 pounds) is supported by the rim of the wheel, and not by the bearings, no additional friction is caused by its application to a machine. During a visit to the Agricultural Hall, previous to the long-distance championship race [1–7 September 1879] I was surprised to see Cann [one of the contestants from Sheffield] with hardly any preliminary 'spurt' ride more than $\frac{3}{4}$ of a lap with his feet off the treadles, part of this being up the incline. Without the accelerator this feat would be impossible. The theory of this apparatus is somewhat complex, the acting forces being the resultants of centrifugal force, developed by the rapid rotation of the radials, and the vertical attraction of gravity. The action is therefore somewhat on the principle of the gyroscope, for the relative proportions between the weight and the radial distance of the masses is so regulated that a speed is soon attained at which they are virtually self-supporting, and their gravity (or weight) is neutralised by the upward and forward tendency of the centrifugal forces. The additional weight of the accelerator need be no obstacle to its use ...

As a matter of interest, the winner of the six-day cycling championship was a contestant from Newcastle-upon-Tyne, G W Waller, who covered 1404 miles and 6 laps (1 mile − $7\frac{1}{2}$ laps) riding 18 hours a day. Cann, referred to above, came in fourth, having covered 1100 miles and 1 lap. This raises an important question about bicycle design. Should the wheel rims be made heavier in order to enhance the energy storing 'flywheel effect', or could there be a safety hazard in the correspondingly enhanced 'gyroscope effect' making steering difficult? The article above refers to the accelerator preventing swerving on travelling down hill. It would perhaps be wiser not to risk it, especially down steep hills with twisting roads and cliffs on one side.

Another idea that has intrigued cycle designers is the 'unicycle' or 'monocycle'. An enormous 'one-wheel' cycle was described in the *Scientific American* in 1898 as the invention of Vernon D Venable of Farmville, Virginia. The illustration shows a rather grim-faced individual bowling along the road obviously wondering whether it was safer to keep going than to try to stop it without brakes.

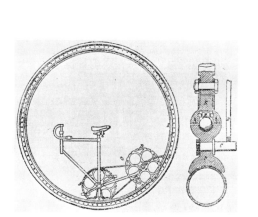

Venable's unicycle, 1898

Another machine totally devoid of brakes was patented by T H Burnes of 7 Bridge Street, Christchurch, Hampshire in 1899. This represented a completely new departure in that it had two driving wheels and like the 'Phantom' bicycle of 1870 it was steered from a central pivot. It also anticipated by half a century the use of small wheels with fat pneumatic tyres.

The frame is a triangle, of which the crank-box is the base. The handlebar stem HS passes down the front tube FT of the frame, into the crank-box where it terminates in a toothed segment, K which gears into corresponding teeth upon the fork head T. A similar stem RS to steer the hind wheel passes down the back tube of frame. Both stems have gear teeth G and G' at their upper ends, which mesh together and are within the head of the frame and cause both to move simultaneously. Thus the movement of the handlebar steers both wheels.

The machine is driven by bevel gear, the lay shafts being in the forked tubes. The steering of the bicycle causes a slight relative movement in the bevel gear wheels; but the angles of steering are so designed that the running is not thereby effected.

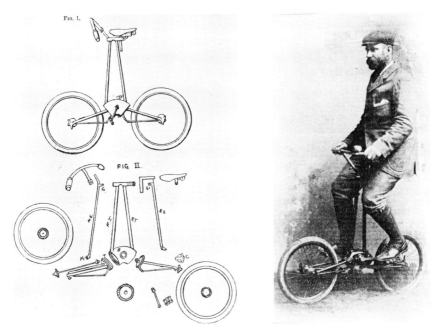

Burnes' double-driver bicycle, 1899

An unconventional machine of a different kind was devised by W G Bullen of 17 Thorncombe Road, East Dulwich in 1898 and put on the market as the 'Orthobaton' bicycle by Messrs Clarke & Co. of South East London. As the illustration shows, the normal handlebar

Orthobaton bicycle, 1898

has been eliminated and the steering is carried out by a twist grip on either side of the seat pillar. This gives the rider a rather unsafe-looking upright posture with nothing to support him in front. It was claimed that the arrangement gave added power in riding against the wind.

A very neat and inexpensive gadget (price 7s 6d (37½p)) obtainable in 1897 from the Walshaw Cyclists' Backrest Co., of Otley, Yorkshire was a strap fitting across the shoulders and hooked on to the handlebars. It was claimed that

> ... it enables the rider to sit upright, or to lean back and rest the back and shoulders while riding, and to gain double the ordinary driving power by bringing the full weight of the body as well as the muscular force of the back and shoulders to bear upon the pedals. The advantage of this is self evident. Everyone is aware of the enormously increased push which can be obtained with the feet without any increased exertion when there is something behind against which to support the shoulders. By the use of the backrest, hills can be climbed with much greater ease than under ordinary circumstances, and with the hands resting lightly on the handlebars, the usual pull on the hands and arms being transferred to the back and shoulders where it becomes an advantage rather than a drawback.

Walshaw cyclists' backrest, 1897

All these machines have one thing in common. They all claim to offer more speed for less exertion. A correspondent to the *English*

Mechanic in 1883, however, was convinced that he had found the final solution—clockwork.

It would of course be a most desirable achievement if there were produced a sufficiently light apparatus which should form part of a tricycle and be capable of its own force of either propelling the machine up hill or materially aiding the rider when making a steep ascent. Subject to the consideration of weight, there is no doubt whatever that such a desideratum could be accomplished by the proper adaptation to the tricycle's axle or otherwise of a powerful spring coil, which upon being wound up by suitable appliances, and being kept wound up in reserve for the occasion requiring the aid, should upon being released so act that its uncoiling would cause the driving wheels to revolve. Some time ago, a successful attempt was made to obtain a motive power of that description with the ultimate view to the propulsion of tramcars thereby. Large and powerful spring coils of a very tough material were made for the purpose, and they were to be wound up by a stationary steam engine at intervals along the route to be traversed by the tramcar .... The idea, however, in connection with tricycles seems sufficiently feasible if only some enterprising machinist would turn it into account. If it should add some 20 or 30 pounds to the weight of a tricycle, the addition would be of little consequence, provided it relieved the driving when ascending steep gradients, for the propulsion of a tricycle along the level or up a slight incline needs but little labour. Supposing the apparatus could be so applied that it should be wound up by the revolution of the machine's axle when going downhill; and even if it involved the necessity of its being wound up by the rider at the foot of every steep ascent, yet would the saving in muscle and backbone power on the rider's part be considerable.

Irrespective of the level at which a tricyclist over ordinary roads in a country not mountainous begins and ends a journey, it may be taken that the total of his ascents upon that journey will, in vertical measurement, be exactly equal to the total of the descents; so that if he travels, say, 40 miles in a day, then 20 miles of the distance will be uphill and 20 down. As the downhill part of the journey needs but little exertion, it is plain that a very large proportion of the tricyclist's labour up those 20 miles of ascent would be saved if only some apparatus were devised that should afford material aid in propelling the machine uphill.

If only ... indeed. This proposal is significant in several ways. Firstly it shows the general desire for some means of relieving the considerable muscular exertions needed to propel a bicycle or a tricycle over the awful roads of the time. Carl Benz, as we have seen, came up with the ideal solution. Secondly it shows complete ignorance of the 'conservation of energy' principle and the connection between work and energy. It was assumed, for example that the spring would take hardly any effort to wind up, yet it would store sufficient energy to carry its own weight plus that of the machine and the rider (albeit with pedal assistance) up any steep hill likely to be encountered. It would be an interesting mathematical exercise to work out what weight of spring (certainly a great deal more than the 20 or 30 pounds that the above writer supposed) would be needed to propel a man and his tricycle on the level—never mind uphill—at, say, 5 miles an hour. A typical tricycle in the 1880s would have weighed in the region of 80 pounds (say 35 kg) and the rider might have added twice that amount. In addition it would be necessary to allow for the weight of all the gearing, both for winding up the spring and for driving the tricycle. A reduction gear of some kind would be needed for winding up the spring in order to keep the effort required within reasonable bounds. An important consideration here is that a very low gear ratio would mean that the time needed to wind up the spring might be significantly long, and this would have to be repeated on every occasion that the spring needed winding up.

Despite the impracticability of the idea of spring-powered road vehicles, a clockwork omnibus is alleged to have run in New Orleans in 1870 but it was far too cumbersome to be viable. How it was wound up is not clear. In 1896, just to prove that 'good ideas' die hard, a coal merchant named W Owen asked in the *English Mechanic* for further information about spring powered tricycles he had heard of that would run for nine hours and only took one hour to wind up. He proposed to use one of these machines for delivering coal to his customers.

A report in the *English Mechanic* in 1879 pointed out that 'although the bicycle has no doubt reached a stage which may fairly be described as an approach to perfection' there was a need for a more convenient vehicle suitable for 'middle-aged riders who not only wish to avoid the risk of a tumble, but who desire to stop for an occasional rest without the necessity of dismounting'. The report discussed at length various improvements in the construction of bicycles and velocipedes generally, in the design and structure of light yet strong wheels, and in methods of lubrication. In particular it gave an illustration of a three-wheeler designed by Mr B A Joule of Sale, Cheshire (not to be confused with James Prescott Joule) with the above requirements for 'middle-aged riders' in mind. The single rear wheel was the driver, and the two front wheels ran on the ends of a centre-pivoted axle. There convention ended. The machine was propelled by the arms and hands, and steered by the feet! Not only that, but the cranks keyed to the rear wheel axle, instead of being at 180 degrees to each other as one might expect, were at zero degrees. That is to say they were both pointing the same way and rotated as though there was only one crank, not two. These cranks were connected by long rods on each side of the wheel to the upper ends of a pair of vertical levers which, at their lower ends, were pivoted on the frame just above the front axle. The rider rocked these levers to and fro by alternately pushing and pulling a hand rail, rather like rowing a boat. Another pair of vertical levers, pivoted to the frame at a point about one third of their length up from their lower ends were operated by the rider's feet. The lower ends of these 'foot levers' or treadles were connected by cords to the front axle so that, for example, pushing the right foot forward pulled the right-hand side front wheel back towards the rider, and the left-hand wheel moved ahead, turning the whole vehicle to the right, and *vice versa*—exactly the opposite to what one might expect. In a modern car this would be equivalent to turning the steering wheel to the left, and the car going to the right.

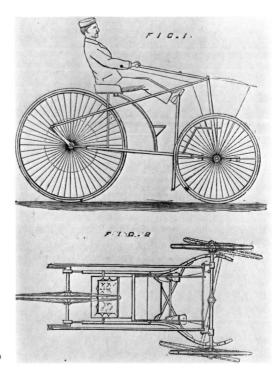

Joule's tricycle for middle-aged riders, 1879

One good reason for having velocipedes with *three* wheels rather than four, was that of having to allow the 'outside wheels' to go faster than the 'inside wheels' on a curve or when turning a corner. This was no problem if the wheels rotated freely on the axles, but if a pair of wheels were keyed to a common axle in order to propel the vehicle then there was a difficulty. Having a single driving wheel avoided that, and it made sense to have the driver on the longitudinal axis to avoid unequal strains on the framework. Despite the differential gear, or 'compensating' gear as it was often called, having been known for centuries, it doesn't seem to have been applied to vehicles to any great extent until J K Starley re-invented it in 1872 (a notable exception being Pecqueur's differential gear used on a steam-powered wagon in 1828). One way of having a double drive, but still allowing one wheel to go faster or slower as the case may be, was to have one or both driving wheels operated by a ratchet gear, or some form of clutch which allowed the 'outside' wheel to be disengaged.

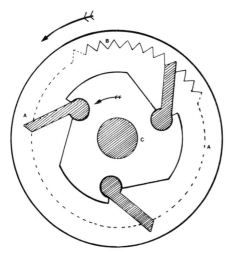

Ratchet drive for velocipedes, 1879

The simplest arrangement was to insert a ratchet gear in the naves of the driving wheels, the dogs or pawls being carried by a plate rotated by the axle, and the teeth of the gear forming part of the nave and hence part of the wheel. The main disadvantage with this system was that going downhill, the wheels could rotate faster than the axle and override the gear. The operator could not then control the speed with the pedals and had to rely on rather crude and inadequate brakes. A common form of brake on velocipedes was a simple lever with a spoon-shaped end which was pressed onto the road surface, which tended to wear it away to a lethal razor-edge. It was also dangerous in that it tended to tip up the machine by lifting one wheel off the ground.

A brake for bicycles that was claimed by its inventor—a Mr Cole—to be 'the only brake that renders it impossible to lose control over the machine under any conceivable circumstance' was put on the market in 1899 by W E Brough of Basford, Nottingham (the father of George Brough of 'Brough Superior' motorcycle fame who had a fatal connection with T E Lawrence who was killed riding a Brough Superior in 1935). In contrast with 'driving by the seat of one's pants', this was braking by the same agency. The rider was required to slide slightly back in the saddle to put his or her weight onto a spring-loaded extension that applied a metal 'spoon' directly onto the back tyre. For an emergency stop the rider just bounced on the brake.

Cole's safety saddle-brake, 1899

Anyone buying an 'Excelsior' tricycle in 1869 might have appreciated a Cole saddle-brake, had it been available. This machine, in an attempt to get the last possible bit of energy out of the rider, was operated by hand and foot simultaneously. The handlebar had cranks exactly matching the pedal operated cranks on the axle of the front wheel. These were linked together by a chain and sprocket arrangement, one on each side. The hands were thus fully occupied going up and down in synchronisation with the feet.

A useful article about velocipedes and their construction appeared in the *English Mechanic* in 1869, written by someone who identified himself only as 'A Member of the High Peak Velocipede Club' from Derbyshire. The article makes interesting reading; it could have something to say to the thousands of joggers who think exercise is a good thing.

Amongst machines of recent introduction, none promises to secure so much health, strength and amusement to the youth of this country as the modern velocipede. It is already being patronised by all ranks; it is not too expensive for the purse of the peasant, nor too insignificant for the attention of the prince. [HRH Prince Albert, the Prince Consort had one. This still exists and can be seen in the Science Museum, South Kensington, London.]

Hygienic reformers are already hailing it as a mode of exercise far superior to cricket, boating, or any other with which we are already acquainted. In the working of a properly constructed velocipede, both the brains and every muscle of the body are brought into use—not at all severe—but (except in racing) in a manner so gentle and healthful that the exercise tends rather to strengthen and exhilarate than to fatigue. Velocipede exercise is especially beneficial to those whose occupations and circumstances compel them to spend the greater portion of their time indoors. If clerks and shopkeepers were to take a six-mile run on a velocipede daily, they would derive an immense benefit thereby. Also ladies who have not sufficient exercise would (with a few months' regular and moderate running on suitable machines) become strong and robust.

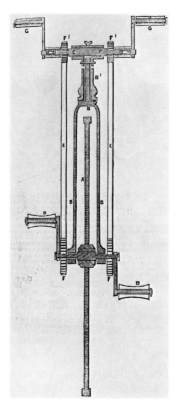

Excelsior velocipede, 1869

In walking a great portion of a man's power is spent in supporting his own weight, carrying; but when he mounts a machine and adopts a rolling motion he has no weight to support, and his whole energy can be exerted in propelling. Walking *versus* travelling on a velocipede is merely carrying *versus* propelling; and it is well known that any animal will propel five times as much as it can carry, on a moderately level road. As it is as important to know what to avoid as well as what to adopt, I will begin with a few points which, to ensure success in velocipede construction, must be avoided.

The main evils to be avoided are weight, friction and speed. As regards weight, no velocipede ought to weigh more than one-third the load which it has to carry, every needless ounce of material tells on the power required. As to friction, the fewer joints the better, and the more direct the power is applied, so much easier will be the propulsion. No combination of cog-wheels or other machinery will really increase power, on the contrary, such superfluities consume a great amount of it. Now as regards speed I would say avoid trying to obtain it by large driving-wheels; of course it would be very easy to show on paper, as is often done, that wheels 5 feet in diameter, at eighty revolutions per minute, will give a speed of fourteen miles per hour; but the power required for such a rate of progression on common roads is far greater than any human being possesses. It will be found far better to be content with small driving-wheels so that a velocipede journey can be performed with ease and comfort, and without any fear of injury to health through over-exertion. The size of the drivers ought in all cases be proportioned to the strength of the 'velocipedist' and the state of the roads. Drivers of 12 or 15 inches are ample for invalids, 2 feet 6 inches for ordinary use (2 feet for hilly districts) and 3 feet, or a little over, for racing purposes—that is, when the races are short. If worked by levers, the sweep of the hands and feet ought to be as great as consistent with comfort, short strokes being as bad as large wheels. The want of this knowledge is the cause of a great many abortive attempts and miserable failures in velocipede construction.

There is at present a great variety of machines by various makers; but after a great number of experiments I have come to the conclusion that the plan adopted by one of the members of our High Peak Velocipede Club is the simplest, safest, and lightest for carrying two persons.

As will be seen from the engraving, it is worked by both hands and feet—the operators sitting face to face and using the same levers.

The two 5 inch cranks are placed at quarter circle, which does away with dead points. A pair of self-acting face ratchets are placed at the inside of each nave, allowing the driving-wheels to turn corners easily, and the levers to be at rest in descending hills.

The wheels are of wood, the rim being in one piece; they are 2 feet 6 inches diameter which is ample for countries like the Peaks of Derbyshire. Spiral steel springs are used, two to each axle; they are light and elastic. For seats, $\frac{1}{2}$ inch deal boards are used which give additional elasticity.

The two gentlemen who have constructed it travel to business on it daily three miles each way, on a very hilly road, the rise being about 300 feet in going, and 180 feet in returning. The time occupied in going one journey ranges from twenty to thirty minutes, according to the state of the roads.

The machine weighs $\frac{3}{4}$ cwt [84 pounds]. It can be taken to pieces in a few minutes and packed for railway travelling in case of a break in the weather. It has been constantly used for nearly three years, so that it has been thoroughly tested.

If the velocipede continues to gain popularity, as it has lately done in

England, both this and succeeding generations will on the score of health reap a most decided benefit—our countrymen and countrywomen will become possessed of clearer heads and more vigorous frames, doctors' bills and doctors' pills will become things of the past . . . .

High Peak Velocipede Club's machine, 1869

Despite giving such detailed information about the machine, the author of the article neglected to say how the velocipede was steered. The engraving doesn't help either, since that part is not shown in the illustration. Neither does he say anything about the brakes; if I were the front-seat rider, travelling backwards, I would want to be reassured about both, especially when free-wheeling down hill. One significant aspect is the use of coil springs for the suspension—a system now much favoured on modern cars. However, I confess to being not quite convinced about the elasticity of half-inch deal boards for sitting on.

The author also mentions friction but doesn't say anything about how to minimise it, except by reducing the number of moving parts. Some considerable thought was given to this matter and the invention of the ball-bearing was a great step forward. However, that didn't meet with universal approval. A correspondent to the *English Mechanic* in 1897 by the name of F W Lane of 87 Cann Hall Road, Leytonstone had this to say on the matter:

Ball-bearings outside a glass case are a delusion and a snare. Their chief use is to be expensive. They are never really lubricated, are difficult to clean, and always requiring it. There is more grinding of grit in a ball-bearing than in the wheel of a dust-cart. The brilliant tests which ball-bearings have passed in triumph have been made under conditions never present in use. Substitute two plain cones for the present cup and cone, and send the balls to the Dervishes . . .

On the same theme another writer the previous year claimed that his old Safety bicycle was all the better for having no lubrication at all:

My Safety has been ridden fairly hard for three years, as often in winter as in summer, say 6000 miles, and that is below the mark, and has never had one drop of oil applied to it, neither chain nor ball. It runs as freely as anything possibly could, and has never required the slightest adjustment, yet all the bearings are as tight as the day it became mine. To my mind this is entirely due to the fact that there being no excess of oil oozing out of the bearings, no dust is thereby attracted, and consequently no grit finds its way in to wear the balls.
This season I had a machine built for my wife, stipulating that not a drop of oil was to be given by the maker; this machine has run some 300 miles, and as freely as if bathed in oil—more so I believe. It would be interesting to know what some opinions are on this subject, or corroboration from those interested. I would suggest anyone caring to try, to take out bearings from one pedal, clean thoroughly, and try it dry against the other, and I am sure, provided the bearings are well made, not another penny will be spent on oil . . . .

I was almost persuaded by his argument, and wondering why my car required so much of the stuff, when I saw his let-out phrase: 'provided the bearings are well made . . .'. I couldn't help noticing too that not only did he do his experiments with his wife's machine, he was also advocating that someone else should do the 'pedal test'.
In 1898 the *Cycling Gazette* of Sydney, Australia carried a report of a home-made bicycle built by a John Gilguys, a farm inspector on a station near Bourke, New South Wales.

The inventor and builder of this machine, was and is, despite his talent for cycle-making, a poor man. He lives among a wilderness of stones, scrub, and sticks, amid the melancholy bleating of thousands of sheep and has to inspect miles of wire fencing. The machine here depicted was the outcome of his lonely leisure and his inventive ingenuity. It will be observed that the bicycle combines two of the latest notions in cycling mechanics—it is chainless and unpuncturable. Though not constructed of aluminium it is rust-proof and at the same time the expense of enamelling and nickel plating is saved. The frame and wheels are of wood, and the saddle is a soogee bag tied with stringy bark. The stays are of twisted fencing wire, and the ball-bearings and spindles have been superseded by ordinary bolts and nuts. The wheels are warranted not to buckle.

Gilguy's rustless bicycle, 1898

Despite the stifling results of the 1865 Act of Parliament which in effect regarded all powered road vehicles—'locomotive carriages'—as though they were lumbering agricultural machines, there were still some adventurous souls who risked prosecution and the loss of their investment by designing and experimenting with engine-driven velocipedes. Many of course took their ideas no further than to air them for comments as paper and pencil designs. Two scientists, however, who actually turned their ideas into reality were Professors W E Ayrton and J Perry at what was then the Finsbury Technical College.

In 1879 these two gentlemen had succeeded in accurately measuring the ratio of the electromagnetic unit and the electrostatic unit of capacitance and obtained from that the velocity of light as being 2.995×10^{10} centimetres per second. They were also responsible for other developments in the techniques of electrical measurement and in instrumentation. In 1882 they took a standard rear-steering tricycle and fitted it with an electric motor. It was then capable of breaking the speed limit by a substantial margin with a speed of six or eight miles per hour.

The driving wheel is 44 inches in diameter and close to it will be seen a large spur-wheel containing 248 teeth. The motor M is slung from the seat platform, and the armature spindle carries a pinion of 12 teeth, gearing into the spur-wheel—the machine being thus speeded down 20 to 1. Taking 460 revolutions of the driving wheel to cover one mile, it will be found that the motor must make about 1200 revolutions per minute to reach a speed of eight miles an hour. The battery, composed sometimes of Faure cells, sometimes of Sellon–Volckmar, and at times of combinations of the two, is slung from the backbone and axle, and, when fully charged contains a store of energy equal to about two horsepower-hours. The motor, which is rather too large, weighs 45 pounds, and with the battery there is altogether a weight of 150 pounds, sufficient to give a speed of six miles an hour. With an increased weight of accumulators a speed of eight miles has been maintained. The steering handle and brake are shown in their usual positions; but on either side of the rider brackets extend, each carrying a small four-candle incandescent lamp to serve as 'lights' and also to illuminate the ammeter A and the voltmeter V by which the rider can see at any moment the amount of current and the E.M.F. between the terminals of the motor, and thus calculate the horsepower which is being expended in propelling

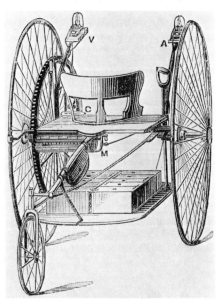

Ayrton and Perry's electric velocipede, 1882

the machine. At the left side of the seat is a commutator C by which the number of accumulators in circuit can be varied, and by which the current can be altogether cut off from the motor. The full power can be obtained only by turning the switch of the commutator through the intermediate powers so that shocks are avoided on starting the machine.

There are two important firsts here. Not only is this the earliest practical electric motorcar in England, it is the first motorcar with incandescent electric lights. Electric lights on cars did not come into general use until after about 1912. The problem with electric motors for cars, then as now, was the weight of the batteries which had to be carried and the limited range available at a 'reasonable' speed on one charge. Because of these seemingly insurmountable problems, inventors turned to other sources of power. Not unnaturally, steam was a prime favourite.

Steam velocipede

An excellent idea for an easy to drive steam-powered velocipede was put forward by an engineer named G J Y Story in 1869. There was nothing remarkable about the engine and boiler, neat and light though they were, except perhaps that the engine had two double-acting oscillating cylinders—a kind often used in paddle-steamers. The important feature which sets this machine apart is that it was gas fired. Coal gas was carried in a concertina-like collapsible holder behind the seat. A weight suspended by a cord over a pulley kept the gas under pressure and forced it to the burner. The state of extension of the concertina showed at once how much gas there was available. The driver of the vehicle had nothing to do but steer and control the speed by a throttle valve on the steam pipe to the engine—and watch out for policemen.

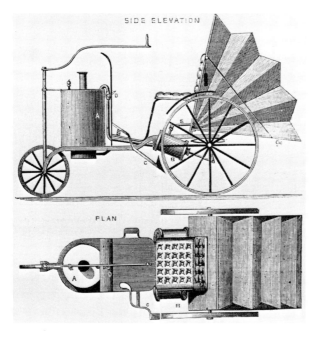

Story's gas-fired steam velocipede, 1869

Another, and in many ways similar, machine was designed in 1870 by a William Stanley of Eyre Street, Chesterfield. This machine was, however, more conventional in that it was coal fired, the driver having to shovel coal from a small bunker in front of him, between his legs under his seat into the firebox behind. Two interesting features about it were that it was fitted with a two-speed gear and it had a pair of buffers at the front in case of accident. More importantly, the front wheels had independent suspension—a very modern innovation. The front wheels were actually carried in bicycle-type forks. A rather frightening arrangement though was that the steering was through a vertical lever on the right-hand side of the seat. This lever had a spring-loaded catch which engaged with slots in a fixed quadrant. The idea was that for going straight ahead, the steering could be set fixed and the driver could take his hands off to attend to the coal and other things.

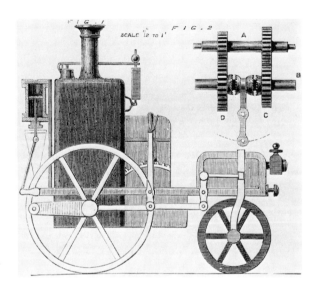

Stanley steam carriage, 1870

The buffers were probably an essential part of the equipment. This machine should *not* be confused with the Stanley steam car built at the turn of the century in the United States by the twin Brothers, F E and F O Stanley.

In 1860 an important step forward (or backwards depending on one's point of view) for personal transport occurred when Etienne Lenoir patented and put on the market the first really practical internal combustion engine. The demand for Lenoir's engine for all kinds of industrial purposes stimulated other engineers to develop much better machines. Space does not allow a full account of the development of the internal combustion petrol (gasoline) engine and in any case that story has already been better told elsewhere. In 1866, the editor of the *English Mechanic* commenting on the state of the coal industry, gave a remarkably far-sighted prediction of the success of the petrol engine when he said:

... Meanwhile those who are engaged in perfecting new motors, who are striving to utilise petroleum as a fuel, may persevere sure of a fitting reward; for one thing at least is certain to rule in their favour, and that is the price of coal. Steadily it is rising in value, and so will their inventions as they are brought out; and they will certainly experience the truth of the old adage, that 'Tis an ill wind blows nobody good'.

A crucial problem with the early internal combustion engines was how to ignite the fuel–air mixture when it was inside the cylinder. Carl Benz adopted electric spark ignition using a Ruhmkorff trembler induction coil and a primary cell to energise it. In theory this should have given no trouble at all, but the primary cell had only a fairly short useful life and since it could not be recharged it had to be periodically replaced; while it was on its last legs the sparks became feebler. One advantage of the trembler coil was that there was no need to worry about ignition timing—it made so many sparks, at least one of them would occur somewhere near the right time. Daimler on the other hand used 'hot tube' ignition. A small platinum tube extended from the cylinder head, the outer end being closed and the inner end being at a point where the fuel–air mixture could enter it as the piston approached top dead centre on the compression stroke. This tube was kept red-hot by a blowlamp. At a certain point the mixture in the tube ignited and the flame then spread to the rest of the charge. This system was extremely reliable—so long as the blowlamp worked properly and the flame played on the tube. It did however have a major disadvantage, and that was it made the engine inflexible. Once started, the engine tended to run at its optimum speed, within fairly narrow limits.

A totally different system of ignition was devised in 1889 by a John Kirkwood of 3 Montgomery Street, Edinburgh.

The sketch inclosed is a novel design of vapour engine for driving a tricycle. Nos. 1 and 2 are working cylinders made of $2\frac{1}{4}$ inch steel tube; No. 3 is an air pump; No. 4 is a very hard fiery grindstone; 5,5 is grooved pulleys for gut band; 6 is an eccentric and connecting-rod with steel pin on end, which touches grindstone at the instant when an explosion is wanted; 7 is an air vessel; 8 is naphtha, which is also connected with air to give it pressure.

Air enters at back of grindstone, naphtha drops on top, and is driven into spray; the naphtha also keeps down dust. Piston passes over exhaust at the end of its out-stroke; the air in air-vessel rushes into cylinder at the same time, and is compressed on the back stroke. The grindstone is the very best exploder, but the most destructive to the engine; but it will run for a time.

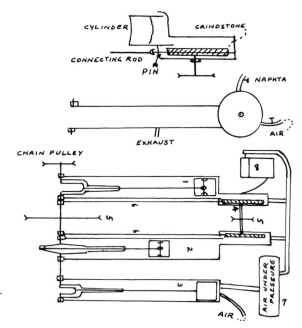

Kirkwood's grindstone-ignition two-stroke engine, 1889

It is unlikely that this engine ever became more than a twinkle in its inventor's eye. It would not have worked satisfactorily, if at all, since the grindstones appeared to be inside the cylinder heads, reducing the compression of the charge to very little above atmospheric pressure. The grindstones themselves wouldn't have been too happy engulfed in flame at every stroke. On top of that there would have been a tremendous problem with grinding wear from the millions of fragments of grindstone trapped in the cylinders; even Mr Kirkwood himself acknowledged that. The fuel used though is interesting, being the lightest and most volatile fraction in the distillation products of crude oil. It would therefore have been highly inflammable and dangerous to carry around.

A slightly better idea was a motorised attachment for a bicycle devised by Nelson S Hopkins of Williamsville, New York in 1896. The engine illustrated weighed only 12 pounds 4 ounces and Mr Hopkins claimed to have built one weighing only $8\frac{1}{2}$ pounds. The engine shown has two cylinders in V formation, the connecting rods acting on the same exposed crank pin. The crankshaft drove the rear wheel through an arrangement of spur gears, without any clutch mechanism. The rear wheel could however be moved slightly so that the cogs came out of engagement—a safety precaution in case the rider fell off. Ignition was by coil and battery.

Hopkins' motorcycle, 1896

In December 1896, shortly after the enactment of the Locomotives on Highways Act, a letter appeared in the *English Mechanic* from F T Reid of 177 Sidwell Street, Exeter with an outline of a proposed electric car, as follows:

I am desirous of making a motor-car to carry two persons—sort of a double tricycle. My ideas were (first thought) of a petroleum motor; (second thought) of an electromotor, and not to require more than 2 volts pressure. From what I understand of electromotor cars, they generally carry motors driven from 50-volt accumulators. I am puzzled to know of any advantage of a motor-car requiring more than two volts when accumulators are carried. It appears to me that to have a motor-car and only one accumulator to work it a vast improvement would be made. (1) There is only one cell and not 25; also by improvement in general, I see no utility in this cell being 25 times as large. The 25 cells must have sides, even if separated by partitions. (2) Each cell must have two half-plates equal to one plate in each cell (inactive) therefore 24 nearly useless plates. (3) Acid much more must be carried than with one large cell, saying nothing about keeping the one cell in good order with ease and economy than 25 cells, one-half of which would be reversing when the other half would be about half-worked out, loss by electrical leakage, &c. I propose using silver wire or bars on armature. Is it 25 per cent better conductor than copper? If so, would it be advisable to use for the first coils F.M.'s silver wire 25 percent smaller to get a greater number of turns nearest the F.M., and outer coils of F.M. copper wire 25 percent larger? Would a four-pole drum be most suitable, or an efficient two-pole machine? I am thinking with so low a voltage the difficulty of getting the amperes through the machine with two brushes only and a conductor (although good) not better than copper.

Lastly, down-hill generally means applying a brake. Well why not utilise this waste of power in running your motor (also a good dynamo) to contribute its energy to the accumulator for the next up-hill? Two volts is baby-work, and could be easily indicated by a galvanometer ammeter, with suitable resistances, reading right and left; current consumed on one side, and current stored on other side which would enable the rider to regulate his machine. How easy to charge this one accumulator in stable: two large cells applied at night, and the result a good spirited horse in the morning willing to do all you require.

Mr Reid put his finger on the fallacy in his argument when he pointed out that 'two volts is baby-work'. To get the equivalent of one horsepower out of his motor, with only a two volt battery, would require a current of 373 amps. A power of the order of 3 horsepower (which would probably be the minimum power required) is therefore going to need a current well in excess of 1000 amps. With 2 volts across the motor terminals and a current of 1000 amps in the circuit, the whole arrangement, according to Ohm's law, could not have a greater resistance than 0.002 ohms. Even without bothering to do the necessary calculations it becomes obvious that this would require not thin silver wire as he was proposing but wire with incredible superconducting qualities. Even if such a wonder conductor existed there is no way that a two-volt motor could be anything but a toy and therefore only capable of 'baby-work' as he so aptly put it. Another very practical consideration is the capacity of the battery to generate such currents at 2 volts. A two-volt accumulator will happily produce a small current for a long time. It follows that if it is required to produce a large current, it will do so only for a very short time and probably be severely damaged in the process.

However unpromising a start this was to Mr Reid's motorcar designing career, the story did have a happy ending. Some how or other he was persuaded that his first thought—a petrol engine—was the best one. In 1900 he sent a photograph to the *English Mechanic* showing the results of his endeavours. It shows Mr Reid with his wife and two children off for a Sunday afternoon outing in their 'pride and joy'. It is obvious from the description which accompanied the photograph that Mr Reid was an adept improviser. His sparking plug, for example, was made from a piece of old barometer tube with a copper wire running through it, and sealed at its outer end. He announced that the car could go at 4 to 6 miles per hour up the steep hills in and around Exeter, with four passengers—and that was no mean achievement. The total weight, without passengers was 728 pounds. It carried $6\frac{1}{2}$ gallons of water to keep the two-cylinder engine cool, and 3 gallons of petrol. The car had two speeds, obtained by a system of flat belts and pulleys and on the level in high 'gear' it could reach ten miles an hour.

Reid's motorcar, 1900

In December 1896, the *English Mechanic* carried the following report about a journey by car from Birmingham to Coventry. It shows very clearly what it was like to ride on a motorcar in those days, and also how limited was the knowledge of how a motorcar worked. Three cars took part, and the people involved were 'journalists and gentlemen'. It is not clear from the report how the journalists and gentlemen were distributed amongst the three cars, but needless to say, the writer was the one on the car that broke down most often. (Notice too that in 1896 one rode 'on' a car, not 'in' it.)

TO COVENTRY ON A MOTOR-CAR

A small party of journalists and gentlemen journeyed from Birmingham to Coventry last Saturday morning. There were three cars, all with Daimler motors, taking part and each was left to make what headway it was capable of. The first to leave was a dogcart built to carry one passenger in addition to the driver. It was the carriage with which the Paris–Marseilles race was won, and being lightly built with a powerful motor for racing purposes, it, of course, covered the distance rapidly and without a hitch. It was computed that the dogcart reached 20 miles an hour, and on declines as much as 30 miles. It was first in and covered the distance [roughly 18–20 miles] in a little over an hour. The second to leave, and also to arrive was a waggonette carrying four persons, and possessing a motor of similar power—six horse—to the dogcart. It was the third in the Paris–Marseilles race. It reached Coventry a few minutes later than the dogcart, without having any 'incidents' to report. The third and last was a Daimler phaeton, except that it had a light covering, built for town work, and propelled by a motor of four horse-power.

The journey through the centre of the city was accomplished safely, though the descent of High Street, Bull Ring and Digbeth raised fears as to the power of the brake, but the tremors were those of inexperience. It was a somewhat difficult task to thread a maze-like path through the traffic which crowded that busy thoroughfare; but the little car, which carried three persons, answered to her steering gear like a smartly built craft. There was a steady run to Coventry Road hill, at the crown of which it stuck fast owing to the carelessness of someone who brought the machine from stabling and neglected to open the oil feed-pipe. This having been attended to, the run to the other side of Hay Mills was accomplished in twenty minutes from town, without deducting anything for the stoppage. The roads were shockingly bad, and the passage of a heavy traction engine and load of trucks—the tracks of which were plainly seen most of the way to Coventry—had not tended to the improvement of the surface. Then it began to rain, and the persistence with which the downpour continued was not productive to cheerfulness. It is, of course, easily understood that when running in the face of a keen breeze, and with a machine made to discharge its fumes underneath the body of the car, there was little of an unpleasant nature to reach the riders, but there were whiffs of burnt petroleum when ascending a hill at slow speed. Nor was the vibration excessive while the car was running; but when standing, the disturbance was so great as to induce the belief that a sufficiently large dose would operate as an effective emetic on a person of bilious tendencies. But this excessive vibration is a defect which can be easily reduced by cutting down the oil supply, and the consequent slowing down of the motor. There was also a jar and jerk somewhat similar to the starting of a cable-car when speed gears were being changed either with the object of reducing or accelerating the rate of travelling. Twice again the hills were troublesome and it was in each instance a sudden sharp incline which caused the stop. With the exception of an occasional tattoo dance by a startled horse, there was nothing further worthy of note until Stonebridge was reached, and pace-making gave way before the temptation of creature comforts. The remainder of the journey was accomplished in very good time. An investigation carried out at Stonebridge had revealed the presence of soot on some mysterious part of the internal arrangements of the motor, and with this removed even Meriden Hill was mounted without any delay. At Coventry the visitors were shown over the company's mill. It had been intended that the journey back to Birmingham should have been by car; but as the rain was still falling heavily, the guests one and all voted in favour of the shelter of a railway compartment. As an instance of the expense of the journey, it may be mentioned that the first car consumed about half a gallon of oil, the cost of which would not be sixpence.

A point to note about that account is that the first two cars at least broke the 12 mph speed limit handsomely. Presumably the 'gentlemen' were sufficiently respected or influential that all the policemen between Birmingham and Coventry turned a blind eye. However, things certainly began to change quite dramatically once it was discovered that the hapless motorist was a prime target and a good source of revenue in the shape of fines for breaches of this or that regulation. It wasn't long either before the police turned to science to trap the erstwhile speedster given to 'furious driving'. Curiously enough it was in France—the one country to take to motoring like a duck to water—where the first attempts were made to devise a gadget that would enable the police to measure the speed at which a motorcar was travelling.

In 1900 a M Delamarre worked out a method of calculating the speed of a moving vehicle using a special camera with two lenses and coupled shutters. The device took a succession of photographs at one-second intervals, and the two lenses having different focal lengths allowed the displacement of the car in a given interval to be calculated from measurement of the displacement of the image on a photographic plate in that interval. A formula was derived to enable this to be done quickly, and the result was independent of the distance of the car from the camera. The only problem was, it wouldn't work unless the car was travelling perpendicular to the lens axes. If the car was travelling parallel to the lens axes, it just generated two superimposed images which were useless for the job. One way to avoid being booked for speeding would be to drive straight at the cameraman.

In the air too things were moving. Ever since the Montgolfier brothers, Joseph and Etienne showed in 1783 that lighter-than-air flight (aerostation) was possible, scientists and engineers bent their minds to devising a means of heavier-than-air flight. The leading member of this movement was undoubtedly Sir George Cayley, a wealthy (and some would say eccentric) Yorkshire Baronet who lived at Brompton Hall, not far from York.

In 1821 Cayley had been instrumental in setting up the Yorkshire Philosophical Society 'to promote science in the district'. Ten years later, following an initiative by Sir David Brewster, the Yorkshire Philosophical Society became a driving force in founding the British Association for the Advancement of Science (originally called the British Association for Men of Science) and Cayley instantly joined as a Life Member. Initially the Association literally *was* for men only. Ladies were not allowed to attend any of the meetings until 1838, and even then they were excluded from anything to do with botany and zoology.

Shortly after the beginning of the nineteenth century, Cayley had begun to publish his ideas about 'aerial navigation' and reports of his experiments, first in *Nicholson's Journal* and later in the *Mechanics' Magazine*. He stated the problem very succinctly when he said: 'The whole problem is confined within these limits, viz., to make a surface support a given weight by the application of power to the air'.

As every scientist and engineer knows, there is more often than not an enormous gulf between the statement of the problem and its solution. Although Cayley laid down the basis for a scientific approach to aeronautics, what was not available to him was a sufficiently light and powerful engine to take him beyond experiments with gliders. However, Cayley's work attracted the attention of other experimenters. In particular it came to the notice of two entrepreneurs in Somerset, William Henson and John Stringfellow.

In 1843 Henson and Stringfellow set up the world's first airline—the 'Aerial Transit Company' and actually tried to persuade people to buy air tickets for flights to India. A piece of verse by an anonymous writer at the time summed up their plans very neatly:

It matters not, I understand, whichever way the wind is,
They'll waft you in a day or so, right bang into the Indies!
Or you may dine in London now, and then if you're romantic,
Just call a ship and take a trip right over the Atlantic.

Which when you think of it, is exactly what you can do today by Concorde.

A lengthy description of their proposed flying machine appeared (on 1 April) in 1843 in the *Illustrated London News*. It was designed to weigh about 3000 pounds and it was thought that a lifting surface of 6000 square feet would be needed to get it off the ground. The power unit was to be a 25–30 horsepower steam engine devised by John Stringfellow. Failing to raise enough funds from public subscription to finance the project, they set about constructing a large working model, which they paid for themselves. One of the first things they did was to design various wing sections and try these out in model form by flying them on strings from Third Class carriages on the Great Western Railway (in those days Third Class carriages were little more than open trucks, and the absence of a roof made them very convenient substitutes for a wind tunnel). However, repeated failures, not to mention mounting expense, caused Henson at least to stop. In 1849 he emigrated to the United States and apparently gave up aeronautics. Stringfellow carried on by himself making aeroplane models and incredibly light steam engines but with no real success. In 1868 he was awarded a prize of £100 for an extremely light steam engine shown at the Aeronautical Society of Great Britain exhibition at the Crystal Palace and he spent the money on continuing his experiments until his death in 1883 at the age of 84.

Over a period of years a Manchester engineer by the name of F D Artingstall carried out a number of experiments with steam powered models and in July 1866 he wrote to the then newly founded Aeronautical Society describing what he had done. His letter was found to be of sufficient interest to be read to the members at their next meeting.

When a young man I witnessed the experiments with the locomotive engines that were tried at Rainhill near Liverpool, and which established the present railway system. Seeing the vast amount of power in those machines, with more zeal than science I thought it would be not very difficult to make an engine to fly by steam.

Mr Artingstall then went on to describe a model flying machine that he had made and tried, not very successfully. He suspended it by a cord from the ceiling. When the steam was turned on the wings, which were designed to flap up and down in imitation of bird flight, just shook themselves violently to pieces and after a few minutes the boiler exploded. Eventually he had a brilliant idea. He recognised that one of the problems was the poor power to weight ratio and his solution was to separate the flying part of the machine from the energy source—the boiler. He did this by joining the flying part to the boiler, which stayed on the floor, by means of a long flexibly jointed brass tube.

When all was ready the generator was put on the fire—the engine and wings at the end of the long pipe rested on a post about two feet from the ground.

I turned on the steam at the generator, when to my great astonishment and satisfaction the engine instantly flew into the air and kept itself up to the end of its tether.

An illustration of an Artingstall machine was published in the *English Mechanic* in May 1868, together with details of the mechanism. The whole machine weighed no more than 15 ounces and the wing span was 64 inches from tip to tip, the wings being 9 inches wide at their widest point. This particular one was intended to be operated by compressed air.

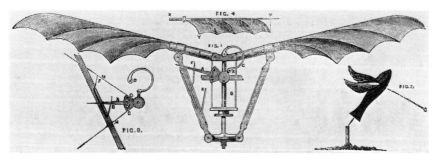

The Artingstall flying engine, 1868

Sadly, Mr Artingstall died suddenly shortly afterwards, leaving his wife and invalid daughter completely penniless. He had spent all the money he had on aeronautical experiments. However, the members of the Aeronautical Society thought so highly of what he had done to further the cause of heavier-than-air flight that they had a collection and presented the widow with a small donation in recognition of his achievements.

In 1866 a patent was granted to a Richard Boyman for his rather unorthodox design for a steerable airship, to be powered by a 406 horsepower reaction turbine. The most interesting thing about this machine however, was the 'balloon'. It was to have been made of thin steel in the form of a cylinder with conical ends. The body of the cylinder was to be 200 feet in diameter and the whole thing was *a quarter of a mile long*, weighing 600 tons. The balloon was to be filled with coal gas, which also doubled as fuel. What would happen as the gas was consumed wasn't specified.

A rather more well-known machine was described in 1867 by its designer, Joseph Meyers Kaufmann of Glasgow. He realised that it was much harder to get a flying machine off the ground and into the air, than it was to keep it there once it was flying (conveniently ignoring any questions about control and stability and so on). In order to help his machine upwards it was fitted with steam-operated telescopic legs. When steam was suddenly turned on, these legs would shoot out and the machine would in effect leap up into the air rather like a gigantic housefly taking off. Playing safe, the machine was not only equipped with wheels for landing on terra firma, it could also float on water. Another interesting idea—which was actually used in the 1940s during the Second World War—was to tow passenger-carrying gliders behind.

Mr Kaufmann presented a paper on his experiments to the Glasgow Mechanics Institute which was reported in the *Mechanics' Magazine* in June 1869. His confidence in his ability to bring his plans to fruition was summed up in his concluding remarks:

My paper has now been long enough to wear out your patience. I will, therefore, conclude, firmly believing (if properly operated) in the impossibility of failure.

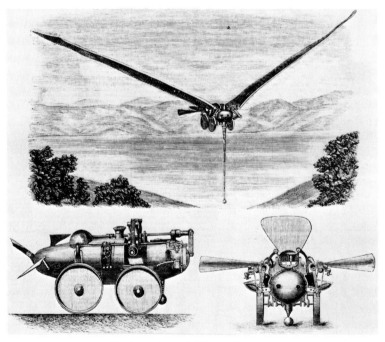

Kaufmann's aerial locomotive, 1869

One thing Kaufmann did do was to build a working model to illustrate the effectiveness of his design. The illustration is taken from a photograph of the model (together with an artist's impression of it flying) which appeared in the *English Mechanic* in March 1868. A description of a trial of this model was as follows:

The model of this machine which was made and tried very recently is of about one horse-power, and weighed (complete) only 42 pounds. It was constructed to prove the practicability of obtaining that indispensable combined oscillating and screwing motion of the wings as attached to the body, and without which the act of flying could not be effected. The engine consists of a direct-acting vertical cylinder, working the wing beams through the cross-head and connecting rods, kept in position by a pair of guides; the main shafts drive the motions which throw the wings in and out of position at each stroke. At the last trial this machine was strapped down to some heavy planking and loaded with a considerable weight to prevent its moving, while the wheels were kept off the ground by means of supports. The experiment lasted but a short while, on account of the great force applied, 150 pounds per square inch on the piston; the wings after some furious strokes gave way to the strain, which was greater than they were designed to resist.... Altogether the trial, as witnessed by a number of engineers was satisfactory; so much so indeed that Mr Kaufmann is now constructing a machine and carriage of a sufficient power to carry several persons over land and water irrespective of wind and weather. This last mentioned machine is a considerable alteration from the first, and its appearance may be looked for at the beginning of May.

Saying that he had only just started on the full-size machine in March, it was optimistic to say the least to expect it to be ready for flying by the beginning of May.

The helicopter principle too was not neglected. Overleaf is a design for a pedal-powered helicopter made of aluminium. It was the brainchild of a Roderick D Hall from Bradford, Yorkshire. He guessed it would weigh no more than 60 pounds.

A variation on the helicopter also shown overleaf was the flying saucer (or flying bath tub depending on your point of view) invented by Estanislao Caballero de los Olivos of 34 West Fifteenth Street, New York City.

Fortunately the bath tub was fitted with springs underneath as a precaution against heavy landings. The idea was that the screws would provide vertical lift up to the desired height, after which the machine would glide, the direction being controlled by the tiltable disc.

A rather better machine was invented by Professor Charles E Hite of the University of Pennsylvania, in 1898. Professor Hite was a naturalist and he had been with Lieutenant Peary in several attempts to reach the North Pole. He came to the conclusion that a flying machine offered the best prospect of success in that kind of venture.

Power was provided by a 15 horsepower engine running on compressed carbon dioxide and motion was provided by two six-bladed

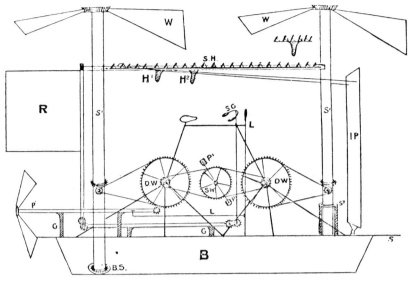

R D Hall's helicopter, 1899

propellers, one on each side of the balloon, driven by belts passing over wooden pulleys. These propellers were carried by steel tubes which formed extensions to the framework of the basket. So confident was Professor Hite that he promptly established the Hite Aerial Navigation Company to benefit from the profits he was going to make.

Caballero de los Olivos' flying machine, 1895

A trial machine was actually built and demonstrated in East Trenton, New Jersey by the New Power Company which was then in Meade Street. The disposition of the propellers though was hardly conducive to successful control of the machine. In anything like a breeze it would have been completely unmanageable. Hite mistakenly assumed that the propellers on each side would operate in air in exactly the same way that side paddles operate on a paddle-ship. In practice, they would have tended to pull each side in opposite directions. At best, the cigar-shaped balloon would have progressed sideways.

Professor Hite's airship, 1898

An unusual machine which actually did fly successfully and was demonstrated in various parts of the United States was the 'Sky Cycle' designed by Carl E Myers of Frankfort, New York. The illustration shows his 1899 man-powered machine which he announced in the *Scientific American* to be the result of ten years of experimenting. It was actually a combination of a balloon and a kite, the base of the balloon being in the shape of a slightly curved plane tilted at a slight angle to the horizontal. The whole machine plus the rider plus 30 pounds of sand for ballast only weighed 236 pounds. The balloon part was inflated with hydrogen but the uplift was obtained by the forward motion of the machine when the rider furiously pedalled the propeller, the whole thing having slight negative buoyancy. Myers claimed to have made flights over the states of

Maine, New Hampshire, Massachusetts, Connecticut, New Jersey, Delaware, Maryland, Virginia, Ohio, Michigan and Illinois and over nearly every county in New York State. He also claimed to have on one occasion reached a height of 10 000 feet, although mostly the flights were at low level. He expressed the view that attaching a suitable engine might improve the performance, but it seems likely that it would have tended to oscillate if the forward speed was significantly increased.

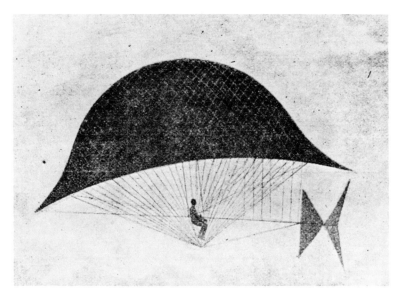

Myers' sky cycle, 1899

Between 1891 and 1896, Professor S P Langley, an American physicist known for his work in connection with thermal radiation (and the invention of the bolometer) spent a great deal of time, effort and money on developing a flyable model aeroplane at the Smithsonian Institution, Washington. Confusingly he insisted upon calling his creation an 'aerodrome'. In 1895 he succeeded in getting his sixth machine to fly and then went on to build a full-size machine. This was launched using 'catapult assistance' from a floating platform on the Potomac river. Unfortunately it was not only aerodynamically unsound, it was also structurally too weak. It plummeted straight into the water. It was, however, fished out and after repairs and modifications he tried again with exactly the same result. Then he gave up.

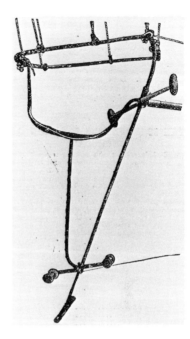

Harness and pedals for Myers' sky cycle

The only successful model, Number 5 (it was the sixth model, but his first one was Number 0), was powered by a small steam engine using a flash steam generator. This was simply two parallel closely wound cylindrical coils of small-bore tube, kept at red heat by a paraffin (kerosene) burner. Water was pumped in continuously at one end and emerged as superheated steam at the other. The single cylinder of the engine was 33 mm bore and 70 mm stroke. At 800 revolutions per minute 0.56 brake horsepower was developed with a steam pressure of 80 pounds per square inch.

In 1897 the *Aeronautical Journal* reported the following account of one of the flights with this machine in 1896, as witnessed by a newspaper reporter:

I saw this machine, made chiefly of steel, weighing as much as a four-year-old boy, yet so large that it would just about fill the average parlour, moved by a steam engine which was part of it, dart forth from the launching stage and fly in an almost straight line through the air a distance of more than 1500 yards, or over three-quarters of a mile. The flight was horizontal. There was not a quiver of the wings and the great bird-like aerodrome swam, as it were, upon the planes of the atmosphere.

This was almost certainly an exaggeration of the aerodrome's capabilities. However, it did fly after a fashion. Langley's later full-size machines were powered by internal combustion not steam, although, as we have seen, the engine was not able to cancel out the aerodynamic inadequacies.

Langley's 'Aerodrome', 1896

An account of the state of aeronautics at the end of the nineteenth century was reported in the *English Mechanic* in August 1899 by F H Wenham (also mentioned in Chapter 2 in connection with experiments in gas lighting). An edited version of Mr Wenham's report is as follows:

I was one of the promoters of the Aeronautical Society of Great Britain, inaugurated on 6th January 1866, and for many years served on the council. We received a number of communications, some of them of such a fantastic and grotesque character as to be beyond the pale of public notice. Some proposed complicated arrangements of levers and cords and internal movements by which the effects of gravity were to be neutralised or inverted by creating a persistent unbalanced force which even these schemers themselves could not verify.

A few attempted to demonstrate the problem by mathematical deductions. Sheets were filled with symbols, starting from a mere assumption, the end culminating in the position of 'as you were' showing that mathematics in such cases is an artificial reasoning that forms no substitute for the inventive brain of practical mechanics. There appears to be a lamentable ignorance of previous experiments with flying machines which have before been found to fail. If such had been studied, with the cause of failure, many abortive schemes would not have been attempted, and so have spared the loss of time and money of inventors. If Mr Maxim, who has built the largest and most costly steam flying machine, consisting of an immense and nearly square *flat* plane driven by screw vane, had looked up and carefully noted precedents, he could have foreseen causes for failure, as similar arrangements had been tried long before. He may have obtained a lifting effect of the whole contrivance; but what of that? For in a brief free flight the machine would have come to grief by reason of improper stability, and the first voyage of any foolhardy persons who ventured out in it would certainly be their last.

It may be asked, What right or reason have I for criticising? I first noticed the question of flight, and began to experiment on the subject in the year 1858. At the first meeting of the Aeronautical Society in January 1866 I gave a *resumé* of these experiments, and the deductions derived therefrom. These appeared in their *Transactions*. I find them also published in book form by Messrs. Cassell and Co. of which I possess one copy, under the title 'Aerial Locomotion'. I did not assume any fanciful movements or actions to explain bird flight, which I accounted for on tangible dynamic experiments and reasoning. To acquire flight with a machine constructed in imitation of a bird's mechanism is utterly impracticable, for such an arrangement can never meet with success. Flapping wings as a means of support for a flying machine is an impossibility; like arrangements were first tried a century ago, and they have all ended in failure. The only feat of bird-flight that approaches the condition of an aerial machine is when birds soar at a level for nearly a mile at a time, with outstretched and motionless pinions. This occurs with the largest and heaviest species, such as vultures, pelicans, the albatross, &c. Yet in the face of all previous knowledge we still have proposers of machines to be supported by vibrating wings; I have stated that the lifting of the down-stroke of a surface is almost nil. Try at first with a lady's fan, say not surface enough, increase this till all your strength is used up in waving it up and down, still no considerable supporting effect. Do not try to imitate nature as in animal progression. If the first pioneers of locomotion, such as Murdoch or Trevithick had done so, they would have produced a hideous multipod affair resembling a huge centipede! Whether the flap of a wing is directly downward, figure of eight, or any other delineation of stroke, no marvelous difference of effect will result. Flight is most likely to be obtained by a *continuous* force.

Of all the numerous models that I have seen during my lifetime, I have observed nothing to surpass in performance the screw-vanes rotated by a spring, or wound string, and set free in air. At one time I tried many of these made of tin-plate for the convenience of altering the pitch and beating the surface into form. These usually flew round with circling movements with bird-like performance, and sometimes went away fairly straight to a distance of more than one hundred yards so far out of sight that I lost several. These screw vanes made out of stamped sheet iron, all to one form, for flying shot practice, are now sold with a coiled spring projector by the leading gun makers in lieu of the hollow glass balls thrown up from a catapult. These screws are remarkable for their stability. From this position they do not deviate till struck by a charge of small shot, when they will sometimes turn a somersalt. In my papers of 1866 I gave an estimate of the approximate force required to prolong the flight of such a screw and found it to be

3 horsepower per 100 pounds raised; but this power consumed is far in excess of the reality, as the maximum of economy is at the first start. Towards the end of the flight when the velocity ceases, there is a waste of lifting power on the ever-yielding air, accompanied by a pitch unsuitable for a decreased velocity. Screw vanes are destined to play a most important part in aerial machines as propellers, and probably in part as lifters; but no model that I have seen has a form of surface adapted for maximum of abutment on elastic air. The surface should have an expanding pitch—that is, an increasing angle in a highly progressive ratio. There is also loss from centrifugal force, tending to drive the air outwards. This must be counteracted by curving the surface of the blade downward, the curvature increasing rapidly towards the extremity. The component of these curves will produce a very hollow surface of what has been termed a 'conchoidal' form. I intend to test this by making a screw in accordance with these conditions. The principle of aerial flight is not a very abstruse one. In this we have the problem of supporting a heavy inelastic body on a medium of light weight and great elasticity. We can only do this by quickly grasping at a large body or wide stratum of the light element before it has time to yield; bringing the law of first impact or undisturbed inertia into action, to serve as a substantial support. A rapid forward movement in still air appears to me to be the only way of successfully effecting this.

All arrangements acting continuously, with no forward progress, set the air in gradual movement without lifting the weight. In fact, as the speed of 'drift' is increased, so is the 'lift' increased by reason of the yield of the stratum of air for support being less. All this I demonstrated by experiment near forty years ago. I carried my experiments to the verge of personal safety. The constructions were deficient in lateral stability, swerving from side to side like an inverted pendulum. I would willingly have continued to experiment, but then had to engage in more substantial business as a mechanical engineer, from which I could spare neither time nor capital.

This is a remarkably far-sighted piece of observation. The problem of stability was not only the most difficult one to solve, it was the one most likely to have been ignored by early experimenters. It was by solving that problem that the Wright brothers, Wilbur and Orville succeeded in obtaining a fully controlled glider flight in 1902 and in the following year were able to make the world's first sustained powered flight. It had taken a century to go from Cayley's first experiments to a successful conclusion. In the succeeding decade the aeroplane was transformed from a primitive 'aerial locomotive' into a sophisticated machine, and even a lethal weapon of war. Once the problem was solved it was easy.

Andrew Scott's prediction

Wha kens perhaps yet but the warld shall see
Thae glorious days when folk shall learn tae flee,
When by the powers o' steam to ony where,
Ships will be biggitt that can sail i' the air
Wi' as great ease as on the waters noo
They sail, an' carry heavy burdens too;
Or, by the powers o' steam yet rise aboon,
An' see what kind o' warld there's i' the moon.

Andrew Scott
A Scottish labourer, 1826